Searching for Minnesota's Native

WILDFLOWERS

Fairy-slipper orchid

ALSO BY PHYLLIS ROOT

Published by the University of Minnesota Press

Big Belching Bog

Illustrations by Betsy Bowen

Plant a Pocket of Prairie

Illustrations by Betsy Bowen

One North Star

Illustrations by Beckie Prange and Betsy Bowen

Searching for Minnesota's Native

WILDFLOWERS

A Guide for Beginners, Botanists, and Everyone in Between

Phyllis Root ✲ Photography by Kelly Povo

University of Minnesota Press
Minneapolis • London

Thank you to all the people, too many to mention, who shared their knowledge about and excitement for native wildflowers with us. A special thank you to Terry Serres for his thorough knowledge, generous help, and entertaining stories. Any mistakes are our own. Thank you, too, to our editor, Erik Anderson, whose enthusiasm and insightful questions helped us find the heart of our book.

To see more photographs and hear more stories of our ongoing journey searching for Minnesota's native wildflowers, please visit us at http://mnnativewildflowers.com.

The University of Minnesota Press gratefully acknowledges the generous assistance provided for the publication of this book by the Margaret W. Harmon Fund.

Published by the University of Minnesota Press
111 Third Avenue South, Suite 290
Minneapolis, MN 55401-2520
http://www.upress.umn.edu

Book design by Brian Donahue / bedesigninc.com
Drawings by Jessica Collette

Printed in Canada on acid-free paper
The University of Minnesota is an equal-opportunity educator and employer.

24 23 22 21 20   10 9 8 7 6 5 4 3 2

Library of Congress Cataloging-in-Publication Data
Root, Phyllis, author. | Povo, Kelly, photographer.
Searching for Minnesota's native wildflowers : a guide for beginners, botanists, and everyone in between / Phyllis Root ; photographs by Kelly Povo.
Minneapolis : University of Minnesota Press, [2018] | Includes bibliographical references and index.
Identifiers: LCCN 2017056070 | ISBN 978-1-5179-0481-4 (hc)
Subjects: LCSH: Wild flowers–Minnesota. | Wild flowers–Minnesota–Identification.
Classification: LCC QK168 .R596 2018 | DDC 582.13–dc23
LC record available at https://lccn.loc.gov/2017056070

This book is dedicated
to all the people
who love native wildflowers
and work to preserve
the places where they
grow and thrive.

Blue flag

CONTENTS

INTRODUCTION

A WHOLE WIDE WORLD WAITING

Once, prairie grasses and flowers grew and bloomed for hundreds of miles in the western part of the place we now call Minnesota.

Once, northern forests of tall white pines and red pines hundreds of years old flourished, and tiny orchids grew among their roots.

Once, big woods of maple and oak and elm sheltered flowers that briefly blossomed every spring.

Once, two friends went looking for these flowers. They looked in places that hadn't yet been plowed or built on or covered up with highways and houses and cities. They wandered through woods, waded rivers, climbed prairie hillsides, and searched in bogs and cemeteries and old railroad rights-of-way and roadside ditches. Along the way, they got lost, wet, itchy, rained on, snowed on, sunburned, tick-bit, and coffee-deprived. But they also saw jeweled shooting star, kittentails, prairie smoke, Dutchman's breeches, twenty-seven kinds of orchids (so far), and more native flowers than they could have imagined. They met people who were generous with their knowledge of plants and places and people whose sense of direction was even worse than theirs. They learned more than one way to say *sarsaparilla.*

They learned that if they followed the clues of time and place and color and leaf, of flower and seed and season and community, they could discover treasures—native flowers that had evolved and grown here for hundreds of years. And, after much looking, they learned that many of these native flowers bloom closer to home as well: in yards and parks and boulevards, along creeks and bike paths and in cracks of sidewalks and corners of convenience marts. They learned that wild Minnesota is sometimes almost right outside the door.

And they fell in love with plants and places.

Now these two friends want to take you on the search along with them. They've already taken care of getting the tick bites, sunburn, poison ivy, and lipping their waders (which is what happens when the creek you are wading turns out to be deeper than your boots are high). Once you finish this book, you just might want to go on a search of your own. They hope you do.

FINDING NATIVE WILDFLOWERS

Searching for Minnesota's Native Wildflowers is a record of our journey of ten years and counting to learn about Minnesota's wealth of native wildflowers by going out to look for them, all around the state and also close to home. Here are some of the ways we hope you'll use this book:

- You can read the book from the first word through the last.
- You can browse the book, reading about specific habitats and flowers that interest you.
- You can find suggestions for places to go to search for native wildflowers.
- You can use the photographs and descriptions of flowers to identify a flower that you find.
- You can read about our searches in the field notes.
- You can start your own list of flowers that you see.

This book has two parts. This first part, "Finding Native Wildflowers," describes what native wildflowers are, why they matter, and how to begin your own search for them. The second part, "A Year of Native Wildflowers," includes descriptions of the flowers we've seen, places to find them, and field notes from our own searches. We organized "A Year of Native Wildflowers" into ten sections, each containing photographs and descriptions of the flowers we discovered in particular habitats throughout the year, moving from spring to fall:

Big Woods
Early Prairie
North Shore
Northern Forest
Minnesota Wetlands
High Summer on the Prairie
10,000 Lakes (More or Less)
Minnesota Peatlands
Autumn Woods
Prairie Fall

The flowers in each section are listed alphabetically by common name. If you don't find a flower in one section, try looking in another section or check the color guide in back—plants don't grow in just one place! Giant Solomon's seal, for example, grows along the North Shore, in the mixed northern forests, in the Big Woods, and in gardens in southern Minnesota. We placed it in the Big Woods section because that's where we first saw it, but you might see it in other locations or habitats as well.

Our photographs and descriptions are intended to get you started. As you get to know more wildflowers, you might want to take along a supplementary field guide for flowers we haven't included or ones we haven't yet seen.

However you use this book, we wish you happy searching.

Dwarf trout lily

Prairie smoke

Pitcher plant

What Treasures Might You Find?

- A tiny trout lily whose only home in the world is three counties in Minnesota.
- A flower so tightly closed that bumblebees must battle their way inside.
- A trillium that blooms so early it often flowers through the snow.
- Cactuses that survive Minnesota's frigid winters.
- Flowers that look like pink smoke across the prairie.
- Plants that trap and eat insects.
- Pitcher plants shining like stained glass in the afternoon light.
- Rare broomrape in a goat prairie where it hasn't been seen for one hundred years.
- Orchids so small you might miss them in the moss under your feet.

Who knows what else you might find? You might even begin to recognize familiar flower faces close to home. There's a whole wild world waiting for you. All you have to do is go look for it.

Prairie, Woods, and Pines

In Minnesota, three great ecological community types (biomes) meet. From the southeastern part of the state a band of hardwood forest made up largely of sugar maple, basswood, and red oak trees (and, once, elm trees before Dutch elm disease arrived) reaches diagonally toward the northwestern corner of the state. On the northeastern side of this band, the northern mixed forest reaches down from the north, made up mostly of pine trees, spruce trees, fir, aspen, and birch trees. And through the southern and western parts of the state, prairie once grew. These three biomes meet in Minnesota as they do nowhere else in the United States, creating many smaller ecological communities. A fourth biome, Aspen Prairie Parkland, dips down from Canada in the far northwestern corner of the state, a landscape woven from the meeting of woods and tallgrass prairie where aspen groves and burr oaks mix with prairie grasses and flowers. Minnesota is rich in diverse landscapes and the plants that grow in them.

Fire and ice formed the landscape where these plants grow. Lava flowed and hardened. Glaciers, great sheets of ice, moved down across the land over thousands of years, melting back and then re-forming as climate warmed and cooled. These ice sheets scraped away soil from some places and dumped it in other places, forming hills and depressions. Blocks of ice trapped in sediment melted as the glaciers retreated and formed lakes and wetlands. More melting ice formed rivers and streams. The last glacial period, which ended more than ten thousand years ago, covered the entire state except for the far southeastern corner. This corner, along with parts of Wisconsin and Iowa, is called the Driftless Area because glacial drift (the sediment and rocks that the ice carried) didn't get left behind. Instead, the Driftless Area has bluffs, valleys, and the closest thing to mountains for this part of Minnesota. All of this happened over great stretches of time, and today we have a vast complex of wet places, wooded places, dry places, lakes, rivers, bluffs, prairies, and peatlands. And in these places native wildflowers grow.

What Is a Native Plant Community?

A native plant is one that has evolved in a particular environment and habitat over thousands of years. It evolved in relationship with other plants, animals, insects, soil, water, wind, rain, and sunlight. Although we don't yet know all the complexities of our native plant communities and while those communities change over time, we do know that no native plant evolved in isolation from the environment in which it lives.

A nonnative or introduced plant, then, is one that evolved in another place, country, or continent with its own unique habitats and has been brought or found its way into a different country from where it evolved. Even though these plants may look like native wildflowers, they are nonnatives that often crowd out native plants.

Communities of native plants are home for all sorts of birds, insects, and other animals. They are places to mate, lay eggs, hatch, eat, and metamorphose. They are stops on the great currents of migration, refuges for those species that are rare and endangered. Some are found in places of immense silence, others are within earshot of freeways or airplane flyways. Some are so far off the beaten track that you need a topographical map or GPS and a canoe to locate them (and a little luck, perhaps, to find your way back—so far, we've always gotten home again).

You might also find native flowers blooming near where you live. In cities and towns, people have planted native flowers in yards and parks and on boulevards. We've seen coneflowers and wild bergamot blooming by a light-rail station, plains prickly pear cactus on a boulevard, and Virginia bluebells growing in cracks in the sidewalk.

Nonnative or introduced species may be blooming as well, even in wilder places. These are plants growing outside the native habitats where they evolved. A pretty flower face in the woods or prairie isn't necessarily a native flower.

What a Native Plant Community Does

A healthy native plant community provides many benefits. What can it do? Here are a few examples:

- store carbon
- conserve water
- filter water
- purify water
- recycle water
- help prevent or moderate flooding
- help prevent erosion
- build soil
- recycle nutrients
- reseed itself
- make oxygen
- provide food for insects, birds, animals, and people
- provide nectar and pollen for pollinators
- provide habitat and home for animals and plants and insects and birds

And . . . provide places for native wildflower seekers to find plants that thrill and delight them.

How Do You Look for Native Wildflowers?

Naturalist and essayist John Burroughs once said that any plant whose name we do not know is a stranger. It is worth getting to know these native wildflowers with whom we share our world and that have been growing here for hundreds or thousands of years.

But how do you know what to look for? How will you know treasure when you see it?

First of all, go slowly. Take your time looking. One of the joys of looking for native flowers is the peaceful pace. When you find a plant you want to identify, look closely and try asking these questions about it. Asking lots of questions is one way we've seen and learned about native wildflowers over the years.

What color is the flower?

Look in the color guide at the back of this book to help narrow your search. But remember: colors also vary. For example, if you don't find the flower you are looking to identify in the red section of the color guide, try the orange section.

Wood lily

Black-eyed Susan

How many petals does the flower have?

Trilliums, for instance, always have three petals. Composite flowers (flowers made up of many individual flowers) often have too many petals to count.

What shape is the flower?

Does the flower have a round, regular shape with three or more petals? Is the flower an unusual shape? Are some petals different from the others? Are the petals fused together instead of being separate from each other? If they are fused into a tube, does the tube point up or down? Is the flower made up of a cluster of many tiny flowers that look like a single bigger flower?

Dutchman's breeches

What are the plant's leaves like?

How do the leaves grow on the stem? Are leaves opposite each other or do they alternate along the stem? Do whorls of leaves circle the stem? Do leaves grow only at the base of the plant (basal leaves)? Are the leaves smooth on the edges or toothed like little saw blades? Rounded or narrow? Do they have small stalks attaching them to the stem, or do they wrap closely around the stem? Do they feel smooth or sandpapery or hairy? Do they have distinctive markings?

Downy rattlesnake-plantain

What is the plant's stem like?

Is the stem smooth or hairy? Round or four-sided? Most members of the mint family, such as wild bergamot, have square-sided stems. Compass plant's stem and leaves are both hairy. Pasqueflower has hairy stems—and hairy leaves and hairy flowers.

Pasqueflower

How tall is the plant?

Some plants are very small, some tower head high or higher. For the ones in between, we carry along a folding ruler to measure plants we can't identify so that we can look them up back home. Starry false Solomon's seal, for example, is shorter than false Solomon's seal, a plant that it closely resembles.

Compass plant

Marsh marigold

Where does the plant grow?

Is it a prairie flower, a woodland flower, a bog plant? Some flowers like damp or shady places, some thrive in dry, open, sunny places. Some plants can adapt to several habitats.

What time of year is it?

Sharp-lobed hepatica

Some plants, such as hepaticas, bloom briefly in the spring, even through the snow, and some only show their colors late in summer or fall.

What does the plant grow with?

What grows in community with the flower you are looking at? You might find red columbine blooming with large-flowered bellwort, or you might see gatherings of asters, sunflowers, and goldenrods all blooming together.

Prairie community

Skunk cabbage

What does the flower smell like?

Great Plains ladies'-tresses flowers smell like almonds. Skunk cabbage is said to smell like rotten meat or skunk, but we haven't yet had the nerve to try to smell a skunk cabbage.

Has the plant formed seeds?

If you see one of last year's plants nearby that looks similar to the plant you're trying to identify, its seeds and seed heads can be a clue. Round-headed bush clover, as its name indicates, has round balls of seeds, as does rattlesnake master.

Even with all of these clues to consider, you still might not be sure what you are looking at. When we first began, we had two large default categories: large yellow flowers and small white flowers. Over the years, we have managed to shrink those categories as we identified more plants, but we still come across "We don't know" plants each year. Even with a good guidebook or website, you might not be able to identify a plant, but it is always fun to try.

Rattlesnake master

Where Do You Look?

When we first started looking, we usually went to places with trails, plant lists, or signs. As we began to learn about what to look for, we headed for wilder places where we could go without paths, not knowing what we would or would not find. We saw some things, we missed some things, we guessed at plenty of flowers, and we learned that we like our wildflowers wild. Over the years we have found ourselves in places with fewer paths, paging through field guides to identify flowers we haven't seen before.

At the end of each habitat section—Big Woods, Early Prairie, North Shore—we listed a few places, some we have been to and some we have yet to visit, to get you started looking for native flowers, both around the state and closer to home (in our case, near the Twin Cities). Most of the places we included aren't quite as wild as the places we go now, but they are great locations for beginning your search for native wildflowers: they usually have paths, sometimes identifying signs, even naturalist tours. It's always a good idea to check with the parks where you are headed to make sure the trails are open. Boardwalks, especially, can have water on them after rain.

Almost any state park will offer the opportunity to see native wildflowers, and many of the parks could be listed for several habitats and for several different times of year. You might come to a park for the woodland ephemerals in spring, return in summer to see a prairie in bloom, then come back again in fall for the blazing color overhead and the bright berries of plants gone to seed at your feet.

Remember, too, that native plants are not orderly. If you don't see a certain flower this week, you can always try again next week—or, if you are too late to see a plant in bloom, try again next year. You can't make a definite appointment with a plant.

Once you feel ready to go on your own without paths, you might consider Minnesota's Scientific

and Natural Areas (SNAs) or Nature Conservancy sites. While rich in wildflowers, these places are often harder to find and explore: they might be wet or steep or rocky; you might encounter poison ivy; and you might get lost without a GPS or a map and compass at the larger sites. If you want to know more about wilder places, we've included the SNA and Nature Conservancy websites in the "Resources" section at the end of this book. Note that some Scientific and Natural Areas are highly sensitive to disturbance and require a permit from the Minnesota Department of Natural Resources to visit.

When you go to places with trails, stay on the trail. Wherever you go, never pick or dig up the flowers and plants that you see. Take a picture,

but leave everything where you find it. It is illegal to take anything out of a state park, and it is also illegal to pick certain flowers no matter where they grow. Be respectful of the places you visit and the things that you see so that everyone will be able to enjoy our native wildflowers.

Go Prepared

We head out with guidebooks and cameras and a sense of adventure—who knows what we'll see?

Depending on where we are headed and what time of year it is, we pack a few other things as well:

- sun hat
- water bottle
- snacks
- notebook and pen
- raincoat if weather threatens
- rubber boots for very wet places
- sunscreen
- insect repellant
- magnifying glass or hand lens for tiny flower identification
- map and compass or GPS (or even all three) for larger places

If you go off the path, as we do, and you are allergic to poison ivy, it's good to wear long pants, socks, and shoes—or even rubber boots. It's also a fine idea to learn how to recognize the three droopy leaves and shiny red stem of poison ivy. Poison ivy thrives at edges of paths and woods, but it also grows exuberantly in the middle of prairies. Poison ivy has made us cautious, but it's never stopped us from visiting sites—we just make sure to go appropriately dressed to avoid a bad case of the itches. Poison ivy has taught us that not all native plants are our friends.

Poison ivy

Wherever we go, we're wary of ticks. We always tuck our pants legs into our socks, and when we get home we do tick checks. So far, the record number of ticks found on one of us is twelve, and we hope never to break that record.

Become a Phenologist

Phenology is the study of when biological events occur in the natural world. For us, it means keeping track of what we see—where and when, whether a plant is in bud, blooming, or gone to seed. We've been doing this for ten years, carrying along a small waterproof notebook and pen or pencil and writing down the places we visit and the flowers we see. Having a record is a great resource, especially if, say, we don't want to miss tuberous grass-pink in bloom or jeweled shooting stars in all their glory.

You might want to keep a list of the plants that you see, where you see them, and when you see them so that you'll have your own record of your wildflower sightings.

Picture This

Native flowers are for looking at, never for picking or digging up. Take a picture instead. You don't have to be a professional photographer to take good flower pictures. The photographs in this book were taken with a 35-millimeter digital camera, and Kelly seldom, if ever, uses a tripod. She doesn't even own a camera vest.

When we started this project, Kelly's goal was to take a portrait of the flowers rather than simply an identification photo. We hope the combination of words and photograph will help you identify the flowers that you find and also tell a story.

Choosing a vantage point is the most important part of letting the flowers tell their story. Try a different point of view. Get up close. Use a shallow depth of field so the background doesn't distract from your subject matter. Always consider the lighting.

Kelly never likes to take pictures on sunny days: the light is harsh and the shadows deep. Most of these photographs were taken on overcast days, in early morning or late afternoon, or, in the case of woodland flowers, in the shade of trees. On a sunny day it might help to have a friend along who will stand where you need them to cast their shadow. (Thanks, Phyllis.)

Basically, if you want to photograph flowers, take a nice 35-millimeter camera, stay in the shade, and get up close and personal.

Click.

A Note about Names

We fell in love with native wildflowers, but we also fell in love with their common names, of which some plants have several. What's not to love about names such as jeweled shooting star, skunk cabbage, bird's foot violet, bluebead lily, rose twisted-stalk, starflower, large-flowered beardtongue, rattlesnake master, rough blazing star, mayapple, kittentails, prairie smoke, and so many more?

Some common names such as greater fringed gentian (which is fringed) or purple pitcher plant (which holds water) describe a characteristic of the flower or plant. Other names such as rattlesnake master (which was said to cure rattlesnake bites, but we don't recommend trying it) describe a plant's supposed abilities. Names such as kittentails, field pussytoes, or prairie smoke evoke images of tails, toes, or drifting smoke. Some names such as Dutchman's breeches, whose flowers are supposed to resemble underwear (breeches) drying on a clothesline, are just plain fun. But why a Dutchman? And why did he need so many pairs of breeches?

Whatever the common name, every wildflower has a specific scientific name that we include in the flower descriptions. That scientific name refers to one particular plant and no other. Whether you call a flower "swamp lousewort" or "swamp betony" or "that twisty yellow flower like a spiral unwinding," when you say *Pedicularis lanceolata*, anyone who knows scientific names will know which plant you mean (although even scientific names can change or vary).

Scientific names are usually from Latin or Greek words. The first word in most scientific names is a noun (sometimes from the name of the person who first identified the plant) while the second word is usually an adjective that modifies that noun. In swamp lousewort's scientific name, *Pedicularis* comes from Latin and means "having to do with lice," from the once-common belief that livestock who ate the plant would get infected with lice. *Lanceolata* is from the Latin word for a spear or a lance and describes a plant with slender, pointed leaves. So, *Pedicularis lanceolata* refers to one and only one specific plant.

You can look at and love flowers without knowing their scientific names (we're proof of that), but scientific names do allow people to communicate about plants very clearly. We still love saying the common names. These flowers, and their names, are poetry, waiting in the world for you to find.

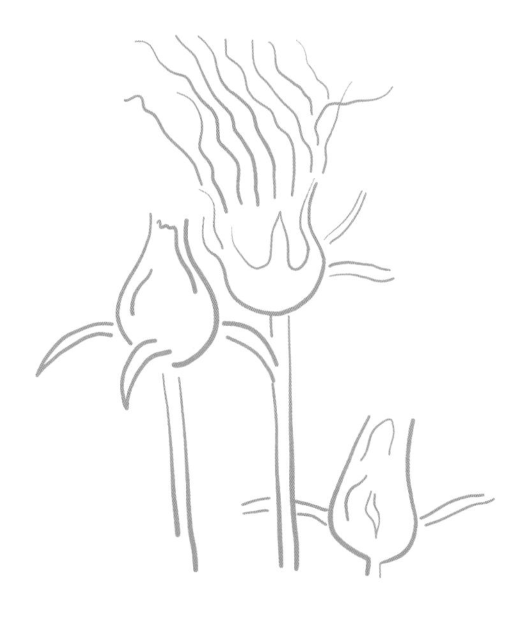

Mayapple

A YEAR OF NATIVE WILDFLOWERS

Each year when snow trilliums and pasqueflowers bravely bloom, we set out to see what native flowers we can find.

Plants don't keep a calendar: they bloom when warmth and moisture and daylight coincide. We've seen pasqueflowers on the same hillside as early as March and as late as the beginning of May. Each year, blooming time can be different from all the previous years.

Neither do plants know where they are "supposed" to be. We've seen small white lady's- slippers in calcareous fens and in wet places on open prairies as well as in roadside ditches. We've seen a single lonely western prairie fringed orchid miles from any other of its kind and a small showy orchis, a woodland species, under a bush alongside open prairie.

We never know exactly what we'll see or where we'll see it. As soon as the snow is off the ground (and sometimes before), we head out to search for this year's flowers, wherever and whenever we might find them.

Here's what a year in search of Minnesota's native wildflowers might look like.

SPRING

BIG WOODS

The Woods Wake Up

The Big Woods was once a much bigger forest than it is now, covering almost two million acres of Minnesota. Maple, basswood, oak, and elm trees grew so thickly that in summer little sunlight reached the forest floor. Early French explorers called it *le bois grand*, the Big Woods. Now only about 5 percent, or roughly a hundred thousand acres, of that original forest remains in fragments that run diagonally through central and southeastern Minnesota.

Flowers that grow in the Big Woods bloom early and quickly. They have only a few fleeting weeks to soak up sunshine before the leaf canopy of the trees gobbles up the sun and shades the ground. Some of these flowers are called spring ephemerals, which means they last only a short time and then disappear completely from sight until the next spring.

The number of flowers considered to be ephemerals seems to change depending on who's defining ephemeral. Although some people use the term "ephemeral" for any early-blooming forest wildflower, true ephemerals vanish completely, leaves and all. At least nine flowers in Minnesota are considered true ephemerals: snow trillium, bloodroot, cutleaf toothwort, Dutchman's breeches, eastern false rue-anemone, dwarf trout lily, white trout lily, yellow trout lily, and Virginia spring beauty.

It has taken us years to see all of the ephemerals. We've waded a river to see dwarf trout lilies, which grow in only a few places in the whole world, all of them here in Minnesota. We've clung to a snow-covered hillside where snow trilliums bloom. We've hiked down dry riverbeds to stumble upon tiny pink blossoms of dainty Virginia spring beauty.

And because ephemerals bloom for only a brief time, we often find ourselves looking too early or too late. But we keep looking. These flowers are too special and too, well, ephemeral to miss.

Spring Ephemerals—Flowers in a Hurry

Bloodroot

Sanguinaria canadensis

Bloodroot's white flowers open on sunny mornings and close at evening, and on a cloudy day they might not open at all. The common name, bloodroot, comes from the red-orange sap in the root that "bleeds" when the root is cut. The single leaf on each bloodroot plant wraps around the flower like a shawl and only opens completely when the flower is done blooming.

Cutleaf toothwort

Cardamine concatenata

The ferny-looking leaves on cutleaf toothwort resemble teeth, but the name actually comes from the toothlike look of its rhizome, an underground rootlike stem. Because the clusters of dainty white or pinkish flowers bloom so early that pollinators may be scarce, the plant has adapted to attract a variety of pollinators, including honeybees, mason bees, cuckoo bees, miner bees, and some butterflies. The deeply cut leaves are so distinctive that we never misidentify them.

Dutchman's breeches

Dicentra cucullaria

Dutchman's breeches gets its common name because the blossoms look like tiny pantaloons, or breeches, drying upside down on a line. The sweet nectar inside the flowers is difficult for most insects to reach. Only those bees with the proper equipment—size, strength, and long tongues such as queen bumblebees have—can fight their way in to reach the nectar.

Dwarf trout lily

Erythronium propullans

Dwarf trout lilies grow only in three counties in Minnesota and nowhere else in the world—our one endemic flower. The flowers of this endangered species measure only about half an inch across. Dwarf trout lilies have four to six white or pinkish petals (actually called tepals), while white trout lilies and yellow trout lilies have six. After years of thinking that smaller white trout lilies might really be dwarf trout lilies, we finally saw the elusive dwarf trout lily, and we knew we would never again mistake anything else for this tiny, fairylike flower.

Eastern false rue-anemone

Enemion biternatum

Oh, the anemones and their look-alikes! Telling them apart has always been daunting for us, but here's what we know: (1) Eastern false rue-anemone flowers (which are not true anemones) almost always have five white petallike sepals, and their deeply lobed compound leaves alternate up the stem. They often form colonies, an early springtime carpet of flowers. (2) Rue-anemone flowers (also not true anemones) all originate from a single point on the stem and have four to nine petallike sepals that can be pink or white. (3) Wood anemone has a single flower that looks white all the way to the center because the stamens are also white, and the flower arises from a whorl of three compound jagged-toothed leaves. Even knowing these facts, we're often confused, but we love seeing any anemones spreading their blossoms over the brown forest floor.

Snow trillium

Trillium nivale

Snow trilliums really do often bloom while snow is still on the ground. (Part of the plant's scientific name, *nivale,* is Latin for snowy.) Each plant has a large three-part green bract that looks like three separate leaves and a single flower with three white petals. Guidebooks say that the flowers are one to two inches across, but when we've found snow trilliums they seem much smaller. Bunches of them can look like a dusting of snow on the ground. Snow trilliums are a species of special concern in Minnesota, which makes finding them all the more thrilling.

Virginia spring beauty

Claytonia virginica

Many of the other ephemerals are white, but Virginia spring beauty brings a burst of color with its tiny pink striped blossoms growing low to the ground. On cloudy days and at night the flowers stay closed and the stems droop, but on sunny days the stems straighten and the buds open. Seeing their striped petals above the narrow basal leaves always makes us think of small peppermint candies. It is a species of special concern in Minnesota.

White trout lily

Erythronium albidum

Trout lilies get their common name because their mottled shiny leaves look like a trout's speckled skin. They are often found along riverbanks and in floodplains. Trout lily flowers face downward with their petals folding back, partly to keep the pollen dry and partly to keep out insects that might steal the nectar. In one hilly riverside area we've seen them growing around the bases of trees like flowered skirts—one naturalist told us of trout lilies he'd seen on a hillside that looked like "river trout running."

Yellow trout lily

Erythronium americanum

You might see whole colonies of white trout lilies in the springtime woods, but we've seldom come across the yellow ones with their rust-colored stamens. Like other trout lilies, yellow trout lilies are often found along streams in wooded places. A trout lily sends up a single leaf each year to gather energy until it has stored enough in its underground corm to produce a flower. A plant can take four or more years to have enough stored energy to flower. So, don't be discouraged if all you see are the leaves; come back next year, and you might be surprised by a yellow blossom.

FIELD NOTES LOOKING FOR SNOW TRILLIUMS

April 1

Last year, a year of no real Minnesota winter, we found snow trilliums in bloom on March 25. This year snow and ice still pile over the ground, but on April 1 we take a chance and head to a steep hillside in Hastings where rare snow trilliums bloom. Snow aplenty, but trilliums nary a one. Too early, too cold, and spring keeps its own calendar. We'll try again in a week or two.

April 11

More snow.

April 18

More snow.

April 26

Ice still lurks in corners from this seemingly endless winter, but the sun shines, and the mercury is headed toward seventy for the first time since last October. Snow is predicted for next week, but today we go looking for spring. What do we find? A tiny snow trillium bud, barely an inch above last year's leaf litter under the trees.

April 28

The snow trilliums have taken advantage of two days of heat and sunshine to open into bright three-petaled flowers. A few hepaticas show purple blossoms, and cutleaf toothwort leaves are popping up. So are the leaves of Dutchman's breeches along with their stems of tiny flower buds like pale green peanuts. These are flowers in a hurry, soaking up sun and warmth before the leaf canopy closes in overhead. We hurry, too, slipping and sliding on the still-snowy hillside, happy to see these first bloomers.

May 3

Snow again.

A Forest Full of Flowers

Canadian wild ginger

Asarum canadense

Canadian wild ginger's purple-brown flowers lie hidden under the softly fuzzy leaves. We loved the story of how flies crawl into these flowers, said to smell like rotten meat, and pollinate them. But it's just that: a story. Canadian wild ginger doesn't smell and is self-pollinating, but it's still fun to peek beneath the leaves to find the dark trumpet-shaped flowers.

Downy yellow violet

Viola pubescens

This one is easy to identify—it's Minnesota's only spring-time woodland yellow violet. The small yellow flowers among heart-shaped leaves have purplish lines on the bottom yellow petal. Short-tongued bees that come to this flower have to turn themselves upside down to reach the nectar.

Drooping trillium

Trillium flexipes

Like other native trilliums in Minnesota, drooping trillium's flower has three white petals and a green bract that looks like a whorl of three leaves. While large-flowered trillium flowers stand tall above the leaflike bract and nodding trillium flowers hang hidden underneath the bract, drooping trillium flowers usually rise above the bract but arch downward on long stalks. The center of the flower can be reddish, the clue that this was a new flower for us—an exciting find.

False Solomon's seal

Maianthemum racemosum

We've often confused the Solomon's seals, and their look-alikes, especially false Solomon's seal and starry false Solomon's seal, both of which have white flowers in bunches at the end of the stem. False Solomon's seal is the one you'll most likely see in the spring woods. It has fewer flowers and is larger and less upright than starry false Solomon's seal, but both are cheery welcomers of spring. You may also find this species referred to as Solomon's plume.

Giant Solomon's seal

Polygonatum biflorum

You might have to look beneath the long arching stem to identify giant Solomon's seal. The small yellow-green tubular flowers hang underneath the stem, usually in pairs, but sometimes in groups of up to ten. Look, too, at the undersides of the leaves, which are smooth, unlike the finely hairy undersides of starry false Solomon's seal leaves—one reason this plant is sometimes called smooth Solomon's seal.

Jack-in-the-pulpit

Arisaema triphyllum

The leaflike spathe (hood-shaped leaf) of a Jack-in-the-pulpit arches over the flower and resembles an old-fashioned canopied pulpit where preachers gave their sermons. This is one of Minnesota's most unusual native plants, easy to identify even though its green and maroon colors blend in with the forest floor. If the plant isn't blooming, look closely at its leaves, whose veins make a distinctive loopy border. You can also spot this long-lived plant easily in the fall by its bright red berry cluster. We never mistake this odd duck of a plant for any other flower.

Jeweled shooting star

Dodecatheon amethystinum

Jeweled shooting star blooms in clusters at the tops of reddish stems that rise from basal leaves. Their light-pink to bright magenta petals gradually fold back as the flower opens. In Minnesota, they grow only in the Driftless Area, and we've only ever seen them at Zumbro Falls Woods Scientific and Natural Area, where clumps of flowers nestle among hardwood trees on top of a limestone bluff. Two-leaved miterwort flowers often grow among them, their tiny white blossoms a delicate contrast to the vivid jeweled shooting stars.

Large-flowered bellwort

Uvularia grandiflora

Large-flowered bellwort leaves look as though the stem has pierced through the base of each leaf. During the two weeks that the flowers bloom, queen bumblebees visit them for nectar. The drooping yellow bell-shaped flowers have twisted petals and may look shy, but they bring an exuberant yellow to the springtime woods. Once, traveling south from Pennington on Scenic Highway we drove for miles past ditches full of large-flowered bellwort glowing yellow in the evening sunlight.

Large-flowered trillium

Trillium grandiflorum

The largest of Minnesota's trillium species, large-flowered trillium plants have a single white flower with three wavy-edged petals. The flowers stand up above the three-part leaflike green bract, and as they get older the blossoms turn from white to pink. Some of the prettiest and most abundant large-flowered trilliums we've ever seen were in a ditch along Scenic Highway—which just shows that you never know where you'll meet a wildflower friend.

Marsh marigold

Caltha palustris

One of the first wetland flowering plants to bloom in spring, marsh marigold's vivid yellow flowers and glossy green heart-shaped leaves shine in the wood's wetter places and in open wetlands. Marsh marigold flowers don't have actual petals, but their yellow sepals look and act like petals. *Caltha*, part of the plant's scientific name, comes from Latin "cup," and the flowers look like little cups, full of sunshine and promise.

Mayapple

Podophyllum peltatum

Mayapple plants, whose white six-to-nine-petaled flowers are hidden under their giant leaves, only bloom after a plant has produced two leaves on a forked stem. If you see a blossom on a one-leafed plant, you can be sure that something has eaten the other leaf. These plants grow in colonies and often look like a congregation of open umbrellas.

Nodding trillium

Trillium cernuum

Trillium comes from the Latin word for three, and trilliums have lots of threes: three white petals, three sepals, and a large green bract that looks like three leaves. The single white flower of a nodding trillium is easy to miss because it hangs beneath the plant's bract, but it's worth looking underneath to see this elegant and exotic-looking flower.

Red columbine

Aquilegia canadensis

Migrating hummingbirds come to red columbine's red-and-yellow-colored flowers for nectar in early spring. So do bumblebee queens and worker bees. The flowers look delicate with their backward curving spurs, but red columbine plants are sturdy adapters—we've seen them in sunny places, shady places, damp places, and even in rocky cracks of sidewalks and alleyways.

Round-lobed hepatica

Anemone americana

Round-lobed hepatica and sharp-lobed hepatica look much alike except for the leaves. Both have three-part leaves, but round-lobed hepatica's leaves are rounder and less pointed. Both hepaticas have hairy leaves and stems, flowers with five-to-twelve petallike sepals ranging in color from white to pale purple, and both are some of the earliest forest flowers. Although the flower bloom times overlap, round-lobed hepatica usually starts blooming in April rather than March, when sharp-lobed hepatica usually starts blooming.

Rue-anemone

Thalictrum thalictroides

The flowers of rue-anemone don't actually have petals. Instead, they have six to ten white or pink showy sepals (the part that is usually green and is found underneath the petals). Rue-anemone's look-alikes, eastern false rue-anemone and wood anemone, both have white flowers with five petallike sepals, although eastern false rue-anemone flowers sometimes have a tinge of pink. Only rue-anemone has up to ten sepals.

Sessile-leaf bellwort

Uvularia sessilifolia

Sessile-leaf bellwort, a relative of large-flowered bellwort, has small, bell-shaped, creamy yellow flowers that hang one to a plant. Unlike its larger relative, large-flowered bellwort, the stems of sessile-leaf bellwort don't pierce through the leaves. If you find one flower blooming, look around—you'll probably find others keeping it company. After years of searching, we were delighted to finally see these small graceful flowers.

Sharp-lobed hepatica

Anemone acutiloba

When hepatica flowers bloom in the spring, new green leaves appear on the plants. In both round-lobed hepatica and sharp-lobed hepatica the leaves turn bronze-colored in the fall and stay on the plants until the flowers start blooming again in the spring. Hepatica is also called liverwort because the leaves are said to look like the lobes of the human liver. We love finding this plant blooming alongside snow trilliums in the very early wet and cold days of spring.

Showy orchis

Galearis spectabilis

This pinkish purple and white orchid is the earliest blooming of Minnesota's forty-nine native species of orchids. Its wide white lip makes a good landing place for queen bumblebees, its main pollinators. While most orchids in Minnesota are more northerly, showy orchis can be found in southern Minnesota. Coming upon this delicate orchid in the springtime woods always feels like a rare gift.

Skunk cabbage

Symplocarpus foetidus

Skunk cabbage likes wet places where it often grows in colonies. It is Minnesota's earliest-blooming wildflower, using starch from its roots to make heat and melt the snow around the plant, keeping an internal temperature of 68 degrees Fahrenheit. The burgundy-to-red spathe part of the flower is shaped like a sharp beak, which helps the plant break through icy crusts of snow. Some researchers believe skunk cabbage plants can live for a thousand years. The second part of the Latin name of this plant, *foetidus*, means stinking, although we have never smelled any skunk cabbages to verify this.

Spreading Jacob's ladder

Polemonium reptans

You'll most likely see the light-blue bell-shaped flowers of this low-growing plant in the woods of southeastern Minnesota as far north as Minneapolis and St. Paul. Flowers grow in clusters from the leaf axils, and the whole plant sprawls slightly and brightly. We've sometimes mistaken spreading Jacob's ladder for Virginia bluebells until we looked at the compound leaves made up of many narrow leaflets.

Twinleaf

Jeffersonia diphylla

It's easy to mistake twinleaf's elegant white eight-petaled flower for bloodroot until you look at the leaves. Bloodroot has a single leaf divided into lobes, while twinleaf's leaf appears to be divided into two identical leaves, almost like a bow tie. This is a flower of special concern in Minnesota; you will only see it in gardens or in the southeastern part of the state, where twinleaf is at the most northwestern part of its native range.

Two-leaved miterwort

Mitella diphylla

The long stem of a two-leaved miterwort has tiny, lacy white flowers that look delicately fringed. Because the plant has such small flowers and two leaves low on the stem, we've sometimes mistaken it for an orchid until we look closer and see the snowflake-like blossoms. Part of the plant's scientific name, Mitella, comes from the seedpods' resemblance to ceremonial hats called miters (the plant is also known as Bishop's cap). You might find whole clumps of these slender spires, on their own or mixed in with other early bloomers.

Virginia bluebells

Mertensia virginica

The petals of Virginia bluebell flowers are fused together in a trumpet shape. The flowers change color as the spring days pass, emerging as pink buds, becoming purple blossoms, then turning the cerulean blue of a spring sky. The flowers grow in masses in the damp woods, often alongside streams, sometimes growing so thickly they look like a flood of blue.

Wild blue phlox

Phlox divaricata

The delicate blue flower clusters of this sticky-stemmed plant are easy to spot in the filtered shade of early springtime woods. The sweet-scented flowers have five notched petals that are joined at the center. Guidebooks say wild blue phlox may form mats of flowering plants, but we've only ever seen them one at a time, standing out brightly blue under the leafing trees.

Wild geranium

Geranium maculatum

Wild geranium is the flower we most often see in the springtime woods—it's almost ubiquitous. Pink to pale purple five-petaled flowers look almost like one-inch-wide rose blossoms in the springtime woods until you look at the plant's large, coarsely lobed, deeply veined leaves. The seed-pods that form after flowering are long and narrow, which might explain another name for this flower: wild cranesbill.

Wood anemone

Anemone quinquefolia

Wood anemone is the only true anemone. Unlike its almost look-alikes rue-anemone and eastern false rue-anemone, wood anemone flowers are completely white, including their stamens. Each wood anemone plant has only one flower, but the plants often spread by rhizomes, making a blanket of white. Anemone is from the Greek word for wind, *anemos,* and anemones are sometimes called windflowers.

FIELD NOTES MINNEHAHA CREEK

We both live in the major metropolitan area of Minneapolis/St. Paul. Cars, highways, houses, and people crowd together. After several springs of driving long distances to see the first wildflowers, we discovered the boardwalk at Minnehaha Park. Here Minnehaha Creek tumbles fifty-three feet in a waterfall and flows on its last stretch to the Mississippi River. We followed old stone steps down, down, down and found where a boardwalk leads under trees and over marshy ground at the foot of steep bluffs. We found other searchers, too, some with their own guidebooks and their children, looking to identify the plants that grow under the trees.

Skunk cabbage grows here, its strange burgundy flowers blooming so early because the plant makes its own heat to melt the snow. Its huge cabbage-like leaves will emerge later when the flowers are done. Not really a cabbage, it's a relative of Jack-in-the-pulpit, which we also find growing close by. Skunk cabbage is said to smell like dead meat or skunk to attract its pollinator flies, but from the boardwalk we can't confirm this, nor do we want to.

Farther down the boardwalk marsh marigold brightens the soggy ground. Trout lily, cutleaf toothwort, and Canadian wild ginger also bloom along the creek. Finally, we climb the steps up, up, up, glad to know there are pockets of woodland wildflowers right here in the city. Who knows where you might find them, once you start looking close to home?

Places to Look

Eloise Butler Wildflower Garden in Golden Valley is thick with blooms in spring, including Dutchman's breeches, Canadian wild ginger, bloodroot, trilliums, Virginia bluebells, large-flowered bellwort, hepaticas, Virginia spring beauty, yellow and white trout lilies, and twinleaf. Especially helpful are the signs telling you what you are seeing and the naturalists who lead regular walks.

Murphy-Hanrehan Park Reserve in Savage has a variety of woodland flowers, including wild geranium and showy orchis.

Beaver Creek Valley State Park in the southeastern corner of the state has wooded trails that burst with Virginia spring beauty, Dutchman's breeches, cutleaf toothwort, bloodroot, large-flowered bellwort, mayapple, Jack-in-the-pulpit, nodding trillium, yellow trout lily, marsh marigold, red columbine, spreading Jacob's ladder, false Solomon's seal, wild geranium, and Canadian wild ginger, among other wildflowers—so many, in fact, that you might want to spend the night camping or staying in a camper cabin.

Carley State Park northeast of Rochester has bluebells, bluebells, and more bluebells—a river of blue flowing across the forest floor. Rue-anemone, eastern false rue-anemone, mayapple, bloodroot, Dutchman's breeches, two-leaved miterwort, large-flowered bellwort, Jack-in-the-pulpit, large-flowered trillium, Canadian wild ginger, and cutleaf toothwort grow here, too. Carley holds a bluebell festival the second Saturday in May, but any day in spring is a great time to visit. Nearby **Whitewater State Park** is also rich in spring wildflowers.

Nerstrand-Big Woods State Park near Northfield is a place we love to visit in early spring. We have seen yellow and white trout lilies, marsh marigold, Virginia spring beauty, trilliums, anemones, hepaticas, Canadian wild ginger, Jack-in-the-pulpit, cutleaf toothwort, wild geranium, Dutchman's breeches, mayapple, and bloodroot. You might even see, in one spot along the park's boardwalk, the rare dwarf trout lily, Minnesota's only endemic flower, growing only in three counties here in our state and nowhere else in the world.

River Bend Nature Center in Faribault has trout lilies, sharp-lobed hepatica, bloodroot, Canadian wild ginger, Dutchman's breeches, Jack-in-the-pulpit, Virginia spring beauty, and mayapple. Come back later in the year to the prairie loop trail to see compass plant, yarrow, round-headed bush clover, black-eyed Susan, butterfly-weed, wild bergamot, leadplant, purple prairie clover, white prairie clover, and big and little bluestem grasses.

Prairie smoke

EARLY PRAIRIE

The Springing of the Prairie

Prairies, those rolling seas of grasses and flowers, once covered eighteen million acres of Minnesota. French explorers called these seemingly endless expanses *les belles prairies,* the beautiful grasslands. We see the prairie aboveground, dense, vivid, and glorious. But the prairie lives below ground as well—up to twenty-four thousand pounds of roots in an acre of native prairie. Some prairie plants, such as big bluestem, grow roots three times longer than the plant is tall. Roots help prairie plants survive drought, drying winds, and fire. They anchor the plants in the soil so tightly that houses were once built of blocks of prairie sod, and wooden plows broke trying to cultivate the ground.

Early settlers thought the prairie might stretch on forever, but now only about 1 percent or roughly 235,000 acres of the original prairie remain in Minnesota. A small fraction of that 1 percent is virgin prairie, never plowed and never grazed. Those remnants still contain an abundance of native plants, from ground-hugging cactuses to big bluestem grasses up to eight feet tall.

Pasqueflower

Kinds of Prairie

Not all Minnesota's prairies are alike. Many forces helped to shape a prairie, including rain, soil, fire, and wind—and once, vast herds of bison. How wet or dry a prairie might be, what kind of soil it has, whether trees grow on the edges—all these make each prairie unique, with its own particular mix of flowers and grasses. Even within a single prairie the ground might be wetter in some places, drier in others. Wet or dry, gravelly, sandy, hilly, rocky, or loamy, all prairies are rich in interconnected species: flowers and grasses

and mammals and reptiles and birds and insects and microorganisms, all depending on each other. In *Grass Roots: The Universe of Home,* Paul Gruchow wrote, "The prairie reminds us that our strength is in our neighbors."

In *wet prairie,* where water is close to the surface of the ground or rain and melted snow drain poorly, you might see more prairie flowers among the grasses. Some flowers that grow well with their feet wet include swamp milkweed, spotted Joe-pye weed, small white lady's-slipper,

swamp lousewort, bottle gentian, great blue lobelia, Culver's root, and nodding ladies'-tresses.

Water drains more quickly in *dry prairie,* sometimes because the soil is sandy or gravelly, sometimes because the prairies grow on hills so steep they are called goat prairies (perhaps because you must be as sure-footed as a goat to climb them). You might see field pussytoes, both white and purple prairie clovers, narrow-leaved purple coneflower, gray-headed coneflower, prairie alumroot, large-flowered beardtongue, plains prickly pear, prairie smoke, and pasqueflower, along with many, many more.

Mesic prairies (mesic means in the middle between wet and dry) often grow on gently rolling ground with loamy soils where rain soaks in. Here you might see leadplant, Sullivant's milkweed, butterfly-weed, Indian paintbrush, asters, sunflowers, lilies, wild quinine, prairie rose, black-eyed Susan, compass plant, goldenrods—the list goes on and on and on. In a remnant of virgin prairie in Minnesota we once identified prairie blazing star, Culver's root, prairie coreopsis, yarrow, leadplant, and rattlesnake master all growing within a three-foot radius, and that doesn't count all the plants we didn't or couldn't yet identify.

A Field Filled with Flowers

Bastard toadflax

Comandra umbellata

Bastard toadflax grows about a foot tall with small white flowers in a tight cluster at the ends of stalks. Before the buds open, they look like bunches of tiny, smooth white balls. The stem and leaves are yellow green. This is a plant you might meet in almost any open or partly open habitat in the spring—dry places, sandy places, oak savannas, open woods. Wherever we find bastard toadflax, we always laugh when we say the name.

Beach heather

Hudsonia tomentosa

You needn't be on a beach to see beach heather, although it does grow there. We've seen it in a sandy prairie, its one-quarter-inch-wide yellow flowers with long feathery stamens crowded on wooly-looking stems covered with even tinier leaves. Count yourself lucky if you do see it—beach heather is a threatened species in Minnesota because of loss of its dry, sandy habitat.

Bird's foot violet

Viola pedata

The leaves of bird's foot violet really do look like a bird's foot, similar to a prairie violet's leaf. The best way to tell bird's foot violets and prairie violets apart is to get up close and personal with the little violet flowers. While prairie violets have little white tufts called beards in the throats of their blossoms, bird's foot violets have orange stamens and no tufts, reminding us of a bird's yellow-orange beak.

Clustered broomrape

Orobanche fasciculata

The stems of this small plant (which grows up to nine inches tall) are mostly underground, so the white tubular flowers that grow singly on red stalks emerge directly out of the ground. The blossoms curve almost like tiny periscopes. No part of broomrape is green, because the plant does not produce its own chlorophyll—it is parasitic and gets all its food from other plants. It is listed as threatened in Minnesota.

Downy painted cup

Castilleja sessiliflora

You might not even notice the flowers of downy painted cup at first because the softly hairy, crowded yellow, green, and white leaves of this plant are so striking. Look closely and you'll see the narrow tubular flowers among the leaves. This is a partially parasitic plant, getting some of its nutrients from other plants such as hairy grama and other prairie grasses. It is listed as threatened in Minnesota.

Edible valerian

Valeriana edulis

Edible valerian's long, narrow leaves are edged with tiny hairs that catch the sun and make the leaves look rimmed in light. The plants grow up to four feet tall, with dense clusters of small white flowers at the top. Edible valerian grows in undisturbed places, but because most of its habitat has disappeared, this species is listed as threatened in Minnesota.

Field pussytoes

Antennaria neglecta

Field pussytoes plants grow close to the ground in clusters of gray-green softly hairy leaves. The clumps of small white flowers at the tops of stems really do look like the paws of tiny cats. Because field pussytoes plants give off a biochemical substance that reduces the height of nearby grasses and flowers, colonies might cover whole patches of ground—a tiny field full of kitten's toes.

Fringed puccoon

Lithospermum incisum

Another name for fringed puccoon is narrow-leaved puccoon. This is one of three species of puccoon growing in Minnesota and the only one with fringed edges to its petals. The flowers glow deep to pale yellow on plants up to a foot tall. Although we may confuse hoary puccoon and hairy puccoon, we never mistake the frilly petals of fringed puccoon.

Golden Alexanders

Zizea aurea

Golden Alexanders flowers remind us of yellow lace with their flat-topped clusters of many small flowers at the tops of stalks. Golden Alexanders plants grow up to two and a half feet tall and might be found in woods or prairies, providing nectar for early pollinators. You might even find Golden Alexanders in gardens, where it prolifically seeds itself.

Ground plum

Astragalus crassicarpus

Small clusters of delicate pink-to-violet flowers bloom on low, sprawling plants. Each flower has a single large upper petal and two small lower petals. What we often notice first, though, are the many narrow compound leaves (made up of many little leaflets) that are hairy underneath. The fruit, when it ripens, is a round purple pod, and the plant is sometimes called buffalo pea.

Hairy puccoon

Lithospermum caroliniense

It's easy to confuse hairy puccoon and hoary puccoon—both have clusters of small bright orange-yellow tube-shaped flowers and hairy stems. But hairy puccoon plants are more imposing—they are taller, the flowers are wider with larger petals, and they often have more flower stems. The flowers are also lighter in color than hoary puccoon. No matter what kind of puccoon we see, they light up the early prairie.

Hoary puccoon

Lithospermum canescens

These are short plants, up to about a foot and a half tall, but they stand out vividly with their clusters of half-inch-wide bright orange-yellow flowers. Turn over one of the narrow-pointed leaflets and the underside will be covered with soft hairs. You might find hoary puccoon, like other puccoons, in dry, open, high-quality prairie where it colors the prairie with its brilliant blooms.

Kittentails

Besseya bullii

The flower spike really does look like a kitten's tail once it grows, but when the buds first emerge from their rosettes of basal leaves, they look almost like a kitten's pink nose. As the flower spike lengthens, small yellow flowers open, a few at a time, from the bottom up. If you find one kittentails plant, look around. You'll most likely find a whole kindle of kitten-tails growing nearby. They are listed as a threatened species in Minnesota—we've only ever seen them on a few dry, open hillsides south of the Twin Cities.

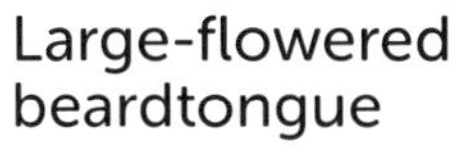

Large-flowered beardtongue

Penstemon grandiflorus

Large-flowered beardtongue's flowers are so big and showy that you might mistake this plant for a garden escapee, as we did when we first saw it. The pale purple tube-shaped, two-inch-long flowers grow in whorls where leaves join the stem, and the plant grows up to five feet tall. Blue-green waxy opposite leaves clasp the stem, making this a plant we can identify even when it's not in bloom.

Long-bracted spiderwort

Tradescantia bracteata

Three kinds of spiderwort grow in Minnesota, but long-bracted spiderwort is the one you'll most likely see. The plants grow in clumps about a foot and a half high with three-petaled, deeply purple flowers with golden yellow stamens. Sepals, flower stems, and long leaflike bracts are covered with furry hairs. Another name for this plant is dayflower because each individual flower only lasts for a single day. We consider ourselves lucky if we visit a prairie on a day when multitudes of long-bracted spiderwort are blooming a brilliant purple.

Pasqueflower

Anemone patens

Pasqueflower buds come up through crumpled baskets of last year's brown pasqueflower leaves and open into delicate purple-to-pale-lavender flowers. Come back in a few weeks to see the silky seed heads on long stems blowing in the breeze. This is the first native prairie flower to bloom in the spring, often through snow, on the south-facing slopes of sandy or gravelly hillsides. Leaves, stems, and blossoms are all covered with fine hairs that help hold heat close to the flower. It is sometimes called "wild crocus."

Prairie alumroot

Heuchera richardsonii

The large, heart-shaped, wavy-edged leaves in prairie alumroot's basal rosette are easy to spot in the early prairie. While prairie alumroot puts up a tall flower spike, the small yellowish-green flowers are inconspicuous. The leaves are showier—they remain on the plant over winter, turning red or brown in fall and regreening in spring to give plants a head start on photosynthesis.

Prairie blue-eyed grass

Sisyrinchium campestre

Even though the leaves on this short plant look like stems of grass, prairie blue-eyed grass is a member of the iris family. Its small pale-blue or white flowers, less than an inch across, open to the sun in the morning and close up at night or in overcast or rainy weather, so even though the plant may be in bloom, you might have to look very closely to see the closed-up flowers.

Prairie buttercup

Ranunculus rhomboideus

This is one of the first prairie bloomers. Glossy yellow flowers may hug the ground when they first open but grow taller over time. The six petals that surround a round green center look as shiny as butter. Minnesota has other buttercup species, but this one belongs only to the prairie.

Prairie smoke

Geum triflorum

Flowers start early in the spring as tight rosy buds on stems above rosettes of dark green leaves and never really look completely open when blooming. The flowers' real glory is the seed heads, which have long, feathery pink plumes that stretch out in the wind and look like smoke drifting across the prairie—and that captivated our hearts the first time we saw them. Like many other spring flowers, prairie smoke's stalks and bracts are very hairy. Bees that pollinate prairie smoke do so by buzz pollination—vibrating their bodies to shake the pollen out of the flower.

Prairie violet

Viola var. *pedatifida*

Like bird's foot violet, the leaves of prairie violet are divided into long, narrow lobes that look very much like a bird's foot. The best way to tell the two apart is to look closely at the bluish-purple flowers. Prairie violet has a "beard" of white hairs in the center, while bird's foot violet has a bright yellow center and no beard. One way to keep the two violets straight is to remember that *prairie* rhymes with *hairy.*

Seneca snakeroot

Polygala senega

These small flowering plants often grow in patches, with pointed spikes of tiny white flowers. Even when blooming, the flowers are round like little buds. At first glance, Seneca snakeroot might look a lot like bastard toadflax, which often grows nearby, but bastard toadflax has one to several round clusters of blossoms on a single plant, while Seneca snakeroot has a single narrow flower spike.

Small white lady's-slipper

Cypripedium candidum

Only a few of Minnesota's native orchids grow on prairies, and small white lady's-slipper is one, blooming in wetter areas (usually of virgin prairie), one flower to a stem, with up to fifty stems growing from a single root. The pouch part of the flower gradually drops down until the flowers looks like little bird's eggs. Small white lady's-slipper is listed as an endangered species in Minnesota.

Starry false Solomon's seal

Maianthemum stellatum

Minnesota has several species of Solomon's seal and its look-alikes, of which starry false Solomon's seal is one. How do you tell them apart? Starry false Solomon's seal, which is smaller than giant Solomon's seal, is the one you'll most likely find in more open places. Its long, arching leaves with parallel veins on an upright stem unfold elegantly into a frond topped by a spike of delicate white starry-looking flowers.

Violet wood sorrel

Oxalis violacea

Violet wood sorrel hugs the ground in dry places and rocky places. The plant only grows a few inches high, but its umbels of small pink to lavender flowers are a cheery find. The leaves are made up of three heart-shaped leaflets that fold in and down at night like an umbrella closing.

Wood betony

Pedicularis canadensis

The crinkly fernlike clusters of wood betony's basal leaves are deep burgundy in spring, with a stout spike at the center. As spring progresses, the leaves turn gray-green and two-lipped yellow flowers emerge sideways, appearing to unwind in a spiral. Another common name for wood betony is Canadian lousewort because people once thought cattle and sheep that ate the plant would become infested with lice.

Yellow star-grass

Hypoxis hirsuta

What look like six yellow petals on yellow star-grass are actually three petals and three yellow tepals that form a six-pointed star at the top of a short stem. Even though its leaves look like grass, yellow star-grass once belonged to the lily family; now it has its own star-grass family. On a wet morning, these small prairie flowers look like a field of fallen stars.

FIELD NOTES SEARCHING FOR PASQUEFLOWERS

March 27

Going to see the pasqueflowers is a yearly rite of spring ever since we read about how they bloom on a hillside along the Cannon River. We always guess at the best time to go looking—too soon and they won't be up, too late and the delicate purple petals are already turning brown.

Always we bundle up and go early (often too early), eager for signs of winter's end. Any time from late March to mid-April, conditions may be right to coax the tiny furry buds out of the ground.

We drive along the river to where a gravelly hillside looms. From the road it's impossible to tell if anything is blooming, but as we climb to the top of the hill we see first one, then several, then bunches of purple flowers spread around in last year's dried grasses like pale eggs in soft brown nests. Prairie children, so the guidebook says, called these flowers little goslings because of the fine hairs that cover stems, leaves, and petals. These hairs help hold in heat, allowing the flowers to bloom even in the snow.

Some years rare kittentails flowers poke their noses out of rosettes of leaves nearby, but this year they aren't up yet. Down along the bike trail under a fallen tree a crowd of bloodroot unwraps its leafy shawls. Here and there prairie buttercups brighten the ground. Clearly, we have not come too early for spring.

Places to Look

Afton State Park on the St. Croix River in Washington County has a restored prairie where you can look for pasqueflower, field pussytoes, golden Alexanders, long-bracted spiderwort, large-flowered beardtongue, and hairy puccoon.

Coldwater Spring near Fort Snelling in Minneapolis has restored prairie where prairie smoke, golden Alexanders, large-flowered beardtongue, long-bracted spiderwort, and many other prairie flowers grow.

Minnesota Landscape Arboretum just west of Chanhassen has pasqueflower, prairie smoke, and other spring wildflowers in its prairies. (The arboretum also has a bog trail and an abundance of wildflowers, both native and nonnative.)

Buffalo River State Park east of Moorhead is adjacent to Bluestem Prairie, a Scientific and Natural Area and Nature Conservancy site. Look for pasqueflower, field pussytoes, bastard toadflax, prairie smoke, prairie alumroot, yellow star-grass, hoary puccoon, starry false Solomon's seal, large-flowered beardtongue, and golden Alexanders. Come back in the fall to find Great Plains ladies'-tresses blooming in the grasses.

Lac Qui Parle State Park near Watson has pasqueflower, prairie blue-eyed grass, golden Alexanders, bird's foot violet, bastard toadflax, small white lady's-slipper, fringed puccoon and hoary puccoon, and many other flowers growing on the prairie hillsides. Great flocks of migrating birds give the park its name, "lake that speaks."

Camden State Park south of Marshall has pasqueflower, prairie smoke, large-flowered beardtongue, prairie blue-eyed grass, and golden Alexanders along the prairie bluff trail. Hike in the wooded valley to see flowers that grow under the trees, such as wild blue phlox.

Quarry Park in St. Cloud has a prairie where you can see field pussytoes, Indian paintbrush, and prairie blue-eyed grass. Walk the wooded paths past old quarry sites and you might see Jack-in-the-pulpit, wild sarsaparilla, giant Solomon's seal, wild geranium, and wood anemone. In the rocky outcroppings, you might even see brittle prickly pear just starting to bud.

You can also visit any of the places listed under High Summer on the Prairie or Prairie Fall to see the first prairie flowers of the season.

SPRING

NORTH SHORE

A Sea of Our Own

Minnesota has ten thousand lakes (depending on who's doing the counting) and one great lake, Lake Superior—the largest freshwater lake by surface area in the world, so big it makes its own weather. Even though the northernmost shore of Lake Superior is in Canada, Minnesotans call the part of their state along Lake Superior the North Shore. Much of that land consists of mixed forests of conifers and hardwood trees such as aspen and birch. But the shore itself, the long, rocky meeting of land and water, is a unique habitat like nowhere else in Minnesota.

Lava flows, glaciers, and time have shaped the North Shore into stretches of basalt and granite over which the waves of Lake Superior wash and where hardy plants find a home. Just inland, some of the flowers you see might resemble their kin growing farther south. For example, panicled bluebells (in the north) and Virginia bluebells (farther south) look similar, but each is a species of its own, formed by climate, water, soil, and sun.

ARCTIC RELICTS
FLOWERS LEFT BEHIND

On Minnesota's North Shore, twenty-three different species of arctic relict plants grow, sometimes hundreds of miles from other plants of their species. One theory suggests that these arctic plants survived at the edges of the ancient glacial ice sheets and were left behind when the glaciers retreated. Another theory is that retreating glaciers left a zone where seeds, carried from farther north, could gain a foothold. However they got here, arctic relicts such as butterwort and bird's-eye primrose are a rare find along Lake Superior's rocky North Shore, growing in microhabitats where the lake keeps temperatures cooler.

Some of these relict plants are so tiny you might never notice them in the cracks and crevices of rock where they grow. Waves crash over them, winter ice scours the rocks, but while these tough, hardy plants persist they can also be crushed by careless feet. Walk carefully and look closely. You are seeing a glimpse of the far north where these plants first evolved.

Northern Beauties

Bird's-eye primrose

Primula mistassinica

Bird's-eye primrose, an arctic relict, grows only about six inches tall on a leafless stem with a rosette of basal leaves. Each flower is about half an inch wide with five blue heart-shaped petals surrounding a bright yellow center. This delicate plant grows mainly along the North Shore of Lake Superior and always makes us happy to find it growing cheerfully in the cracks of North Shore rocks.

Bluebead lily

Clintonia borealis

Bluebead lily is less than a foot tall, with a single stem rising out of a cluster of large, pointed, thick, waxy leaves and with delicate nodding yellow flowers at the top of the stem. You might wonder why they are called bluebead until you see them in the fall with clusters of dark blue berries, which are *not* edible.

Bunchberry

Cornus canadensis

What look like four white petals on a bunchberry plant are actually large bracts surrounding the tiny green flowers at the center of the blossom. These flowers ripen into a cluster of orange-red berries by fall. Bunchberry plants grow low on the ground and have either four or six leaves, but only the six-leaved ones have flowers. Bunchberry flowers are some of the fastest-opening flowers known, even though the flowers are too small to see opening—they've been said to open in less than half a millisecond.

Butterwort

Pinguicula vulgaris

Looking like a star clinging to the rocks, the sticky yellow leaves of this tiny plant attract insects. When an insect is trapped on a leaf, the edges curl in and the plant slowly digests the insect. The small, purple flower blooms on a stalk only a few inches tall. Butterwort is an arctic relict growing in cracks of rock and near tiny pools. If you are looking for butterwort, walk carefully. One wrong footstep can wipe out a whole group of them. Butterwort is a species of special concern in Minnesota because it is so uncommon and needs a highly specific habitat.

Canada mayflower

Maianthemum canadense

Colonies of these small, white spiky flower clusters nestled in wide leaves can carpet a forest floor. Two or three heart-shaped leaves climb up the zig-zag stem (which is less than six inches tall). In the fall, look for bright red berries where the flowers bloomed—but don't eat them. Another common name for this plant is wild lily of the valley.

Harebell

Campanula rotundifolia

These delicate blue bell-like nodding flowers are hardy survivors. They grow along shores and on rocky ledges as well as in dry woods. Like many northern plants, harebells are circumboreal, growing in northern regions around the globe. In Scotland, the very same plant is called bluebells of Scotland.

Panicled bluebells

Mertensia paniculata

The leaves of panicled bluebells, also called northern bluebells, are hairy, unlike the smooth leaves of Virginia bluebells, which grow in the state's more southerly woods. Little clusters of bell-shaped flowers hang from the stems. The flowers are shorter than Virginia bluebell flowers (which remind us of trumpets). One of the common names for panicled bluebells is chiming bells, although we've never heard them ring.

Purple clematis

Clematis occidentalis

Purple clematis flowers grow singly from the bases of the plant's main leaves. What look like petals are really four sepals that surround the flower. The leafstalks twine around nearby plants to hold up the vine, which can grow to six feet long. These large, beautiful purple flowers surprised us along a wooded trail at Gooseberry Falls State Park.

Wild sarsaparilla

Aralia nudicaulis

Whether you pronounce the name *sars-a-pa-rilla* or *sas-pa-rilla,* there's no mistaking this two-foot-tall plant with its large three-part leaves, each of which is divided into three to seven leaflets. Below the leaves, airy globe-shaped clusters of small, white flowers bloom on separate stalks in late spring. Later in summer these flowers ripen into small purple berries. Once the roots were used to give root beer its flavor, but no longer.

FIELD NOTES SECOND SPRING

May 27

Early spring wildflowers have mostly bloomed themselves out in southern Minnesota, but up north they are just beginning—or so we hope. On a rare free weekend, we head north along Lake Superior, intent on finding the elusive butterwort, a tiny arctic relict that grows in the cracks of rocks along the North Shore and is one of Minnesota's four kinds of insect-eating plants.

We're hopeful, too, that we'll see Canada mayflower, bunchberry, bluebead lily, and even an early orchid or two. Will they be done blooming? Will they have even begun?

Rain leads us north, fog blankets the lake, and torrents of little waterfalls pour over the rocks at road's edge. We make an unplanned stop at a state park along the shore to use the bathroom. Curious, we ask a naturalist (who turns out to know some of the same people we do) if any wildflowers are blooming, and even, perhaps, butterwort. She sends us on a trail walk where we find bluebead lily, Canada mayflower, wild sarsaparilla, panicled bluebells, and an unexpected cascade of large purplish flowers we learn are purple clematis, something we've never seen before. A whole second spring surrounds us.

Then the find that propelled our whole trip north: along the rocky shore a helpful ranger points out two tiny colonies of butterwort braving cold temperatures, sparse soil, and splashing waves. We also identify (with some help) Pennsylvania sedge, bird's-eye primrose, and the tiniest violets we've ever seen. In a few days the buds on butterwort will open into vivid purple flowers above buttery yellow leaves, but even when blooming, the plants are so minute, so tucked away under cracks and crevices, that we doubt we ever would have found them on our own.

We love our southern Minnesota spring, but we love this second one, too, miles and hours north. And we've learned that, when in doubt, we should ask people who know more than we do, who have always been happy to help.

Bunchberry

Places to Look

Gooseberry Falls State Park, where we first saw arctic relicts growing, also has Canada mayflower, starflower, bunchberry, wild sarsaparilla, purple clematis, nodding trillium, bluebead lily, panicled bluebells, and Jack-in-the-pulpit, along with rose twisted-stalk and many more.

Split Rock Lighthouse State Park is a good place to find starflower, rose twisted-stalk, bluebead lily, and other northern spring wildflowers. You can see Lake Superior, northern wildflowers, and a historic lighthouse all in one place.

Tettagouche State Park has Canada mayflower, starflower, bunchberry, wild sarsaparilla, nodding trillium, bluebead lily, panicled bluebells, and Jack-in-the-pulpit, along with other native wildflowers, miles of trails, a sixty-foot waterfall, and scenic overlooks of Lake Superior.

Crosby-Manitou State Park and **Cascade River State Park** are great places to hike through the northern woods and look for Canada mayflower, starflower, bunchberry, wild sarsaparilla, bluebead lily, panicled bluebells, rose twisted-stalk, and other North Shore flowering delights.

Our Northern Forest section also includes some of these same flowers.

SPRING

NORTHERN FOREST

Under Tall Trees

The forests in much of northern Minnesota are made up of needle-bearing evergreen trees such as white, red, and jack pine, spruce, and fir, mixed with deciduous trees (ones that lose their leaves) such as aspen, birch, and oak. Scientists call it the Laurentian Mixed Forest. Most Minnesotans call it the north woods or just "up north."

Growing on land that the glaciers scraped over, these forests are dotted with lakes, streams, and rivers. Once almost all of these trees were virgin forest, some of them hundreds of years old and one hundred to two hundred feet tall.

Almost all of the north woods has been logged at least once, but some rare pockets of virgin forest still remain, tree trunks towering and tiny orchids growing underneath. Because needles from pine, fir, and spruce trees along with tamarack needles turn the soil in these forests more acidic, and because some kinds of orchids are often found in mildly acidic soil, one good place to look for orchids is in the neighborhood of pines.

MINNESOTA'S ORCHIDS

Minnesota has forty-nine different species of native wild orchids, from plants with tiny flowers like downy rattlesnake-plantain to showy lady's-slipper, the state flower. Orchids grow in pine forests, prairies, and peaty soil, so northern Minnesota, with its pine forests, bogs, and forested swamps, is a good place to search for them.

Orchids form lifelong partnerships with fungi in the soil, one good reason never to dig up an orchid in the wild and take it home, where it will almost surely die. Some colonies of orchids live for many years if left to grow in their natural habitat.

Discovering an orchid sometimes takes close association with the ground, because many of these plants are so small they blend into the moss or grasses where they grow. But finding either a new-to-us orchid such as a ram's head orchid or an old orchid friend like a small yellow lady's-slipper is always a thrill.

Showy lady's-slipper in bud

Small yellow lady's-slipper

Flowers under the Pines

Gaywings

Polygala paucifolia

These low-growing, shiny-leaved plants have showy purplish flowers whose unusual shape might fool you at first into thinking they are orchids (they fooled us). The unusual flowers have two broad petallike sepals and three other sepals that form a long extension with a fringe at the end. Flowering wintergreen is another name for the plant, but we love gaywings because it makes us think of a brilliantly colored bird in flight.

Goldthread

Coptis trifolia

Goldthread's small flowers look like thready stars, and its three-part glossy leaves with scallop-edged leaflets lie close to the ground. Until we spied this early-blooming plant in a mossy bog and then an old-growth forest, we thought goldthread meant that the plant would be visibly yellow, but the name refers to the threadlike bright yellow underground rhizomes from which the plant spreads.

Lily-leaved twayblade

Liparis liliifolia

Lily-leaved twayblade orchids have two large, wide basal leaves (twayblade means two leaves). The purple-brown flowers, each with a wide lower petal and two spidery side petals, on stalks along a single stem look like those of no other orchid. This orchid is found under oaks and pines, though we've only ever seen it under pines.

Pipsissewa

Chimaphila umbellata

This small plant, also called prince's pine, grows up to ten inches tall and has whorls of shiny dark green leaves low to the ground. The buds look like tiny pink balls dangling on stalks at the top of the plant and open into nodding, downward-pointing, waxy-looking white-and-pink flowers. We've mistaken this plant for an orchid at first (actually, it's a miniature shrub), until we've gotten down to look up at the five-petaled flowers with their prominent green centers.

Rose twisted-stalk

Streptopus lanceolatus

The arching stem of rose twisted-stalk zigzags from long leaf to long leaf, giving the plant its common name. Where leaves join the stem, small bell-shaped white-to-rose-colored flowers hang down. We sometimes confused the white-flowered plants with false Solomon's seal until we learned that false Solomon's seal usually has leaves with wavy edges.

Showy lady's-slipper

Cypripedium reginae

Showy lady's-slipper orchid has been Minnesota's state flower for more than a hundred years (which is also how old some populations of the plants are estimated to be). These eye-catching pristine white and pink orchids can grow up to three feet tall, often in clusters of plants, with one and sometimes two spectacular blossoms per plant in the elegantly unfolding leaves. Part of the scientific name, *reginae,* means queen, and even the unopened buds look regal.

Starflower

Lysimachia borealis

Starflower is one of the few flowers that often has parts in seven—seven petals, seven sepals, seven stamens make up the small star-shaped flowers (although you may also see six-petaled or eight-petaled flowers). Even the pointed basal leaves make a star shape, from which one or two (sometimes three) flower stems rise up. You might find starflower growing in colonies like a starry sky on the green forest floor.

Stemless lady's-slipper

Cypripedium acaule

We first found stemless lady's-slipper (also called pink lady's-slipper or moccasin flower) growing at a campsite in the Boundary Waters when we weren't even looking for flowers. What looks like a stem is actually the flower stalk, while the true stem remains underground. The flower has a deep purplish-pink veined and puckered pouch. Stemless lady's-slipper is one of only twenty-five plants that can grow in the most acidic bogs, but they grow in other places such as woodlands as well.

FIELD NOTES THE LOST FORTY

October 2009

We've come to the Lost Forty, intrigued by the name and the chance to walk in the shadow of what northern Minnesota looked like before the lumber companies cut across the land.

It's easy to see how the forest got "lost," because we get lost trying to find it in the midst of what is now the Chippewa National Forest. The Lost Forty (actually, 144 acres) was surveyed in 1882, when it was mistakenly marked on a map as part of Coddington Lake. Lumber companies back east used the survey maps to determine which sections of land to buy for their timber. But none of the companies wanted to buy a lake—no trees growing there. So, the trees, left alone, just kept growing.

In 1960 a new survey "found" the lost trees, some of them more than three hundred years old. Fire had taken some of the trees, wind and age had toppled others. Many are still so big that two of us can't encircle a trunk with our arms. We tilt our heads back, back, back trying to see the treetops, looking into the sky and into the past.

Lost Forty revisited, May 2016

We've seen the Lost Forty in the fall but never in the spring. The air is fresh with recent rain and birdsong and also mosquitoes (we're glad we brought the bug spray). Even though the pines tower above us, we forget to look up—our eyes are focused downward on Canada mayflower, bunchberry, bluebead lily, and gaywings. Several stemless lady's-slippers grow right along the path. For the first time ever, we see the tiny pink bells of rose twisted-stalk.

Old-growth forest, lost and found.

Places to Look

Minnesota's first and oldest state park, **Itasca,** north of Park Rapids, is rich in habitats and native wildflowers. Along the Dr. Roberts trail from the parking lot to the Old Timer's Cabin we have seen showy lady's-slipper, blue flag, bunchberry, two kinds of trillium, harebell, and red columbine. While you're at the park you can also walk across the Mississippi River at its headwaters.

Jay Cooke State Park south of Duluth has Canada mayflower, wild sarsaparilla, bunchberry, starflower, rose twisted-stalk, bluebead lily, and many more. Take a naturalist-led wildflower walk or hike up to the pioneer cemetery—who knows what flowers you might find?

We went to **Scenic State Park** near Bigfork to see a bog boardwalk, but the park no longer has the boardwalk. The naturalist told us, "The bog ate it." Instead we saw a lovely abundance of springtime flowers, including wood anemone, bunchberry, trilliums, marsh marigold, giant Solomon's seal, starflower, large-flowered bellwort, and showy lady's-slipper getting ready to bloom.

The Lost Forty Scientific and Natural Area, a remnant of virgin forest located in the Chippewa National Forest, has hepaticas, large-flowered bellwort, wood anemone, wild sarsaparilla, bluebead lily, bunchberry, Canada mayflower, rose twisted-stalk, gaywings, starflower, giant Solomon's seal, and stemless lady's-slipper. Follow the paths among tall pines, some of which have been growing for more than three hundred years, and listen to birdsong and wind in the high branches.

SUMMER

MINNESOTA WETLANDS

Wet and Wild

Marshes, fens, and swamps—wet places where, legend has it, swamp monsters dwell and unspeakable things crawl out of the ooze. Wetlands are rich in life, although, as far as we know, no monsters have ever been definitively documented in them. They are places that are shallower than lakes where water can be found at least part of the year.

Wetlands can occur in prairies, in hardwood forests, and in northern forests—anywhere soil, topography, and water keep the ground wet for at least part of the growing season. All wetlands are wet at least some of the time, but not all wetlands are the same. Some are deep, some shallow, some have water rich in minerals, some have water made acidic by peat moss and lack of nutrients, some remain saturated all year, some disappear in summer or during dry spells. The wet places that dot Minnesota may look like nothing more than watery spots on the map, but a variety of native plants and flowers adapted to specific environments grow and bloom in them. Along with coral reefs and tropical rainforests, wetlands are some of the planet's most diverse habitats.

How do you know where you are getting your feet wet?

As a general rule, marshes are rich in minerals. Among cattails, sedges, and reeds you might also find American white waterlily, yellow pondlily, water smartweed, broad-leaved arrowhead, and common bladderwort growing.

Swamps have woody shrubs and trees. The kinds of trees growing in a swamp tell what kind of swamp it is. In a coniferous swamp, where cedar or tamarack trees grow, look for Jack-in-the-pulpit, starflower, bluebead lily, Canada mayflower, and, if you are very lucky, ram's head orchid.

In a hardwood swamp where black ash, birch, aspen, and maples might grow, look for Jack-in-the-pulpit, marsh marigold, skunk cabbage, and, again with luck, purple fringed orchid—luck we haven't had so far, but we keep hoping.

CALCAREOUS FEN
A RARE WETLAND

In calcareous fens, cold water rich in calcium carbonate wells up from underground. Calcareous fens are Minnesota's rarest plant community and perhaps the rarest wetlands in the whole United States. They are found in only ten states in the lower forty-eight states, and two hundred of the known fens are found right here in Minnesota.

Sedges (grasslike plants with triangular stems) predominate in fens, but you will also find calcium-loving plants, called calciphiles, including small white lady's-slippers, a species of special concern in Minnesota, and edible valerian, a threatened species.

Flowers in the Wetter Places

Blue flag

Iris versicolor

In marshes and wet prairies and alongside lakes you might see these two-to-three-foot-tall plants with spectacular blue to blue-violet flowers blooming. Even when the flowers aren't blooming, you might recognize the plant's long swordlike leaves or the seedpod, which is up to two inches long, narrow, and slightly crumpled looking. On our first trip to a bog, we came across a whole field of blue flags blooming, an amazing sight in spite of all the mosquitoes.

Blue vervain

Verbena hastata

The thin blue spires of blue vervain flowers rise out of the tops of tall plants (up to six feet tall) with square stems. Only a few flowers open at a time, starting at the bottom of the flower stalk, then upward along the stalk. This is a plant of wetter places. Other vervains grow in Minnesota, but only blue vervain has this particular lovely shade of blue-violet.

Cross-leaved milkwort

Polygala cruciata

Cross-leaved milkwort is small, with leaves that often occur in whorls of four. What look like white to pink to purple flowers on densely packed flower spikes are really pairs of triangular sepals surrounding the tiny flowers. Cross-leaved milkwort is listed as endangered in Minnesota.

Great blue lobelia

Lobelia siphilitica

Great blue lobelia grows up to three feet tall in wetter parts of prairies, with spikes of deep blue flowers that attract native bumblebees. Each flower is made of two lips, an upper lip split into two pointed parts and a lower lip split into three pointed parts. They are a vivid standout in the grasses.

Lance-leaved violet

Viola lanceolata

Although at first glance these tiny plants look like immature violets, their half-inch-wide flowers are fully grown. Named for the shape of their long, narrow leaves, these flowers are listed as threatened in Minnesota. Most of their original habitat has been lost to development, but some still grow in wet marshes and soggy places. They bloom in the springtime—small but well worth looking for.

Large yellow lady's-slipper

Cypripedium parviflorum var. pubescens

This is Minnesota's (and the country's) most common orchid, one you might see glowing brightly in wet prairies, fens, or roadside ditches. Its yellow blossoms are large, with round pouches and twisting yellow-green sepals with rust-colored stripes and speckles. Small yellow lady's-slippers might grow nearby, but their flowers are much smaller, with dark red sepals.

Nodding ladies'-tresses

Spiranthes cernua

Small flowers twist in a double spiral on the stalks of these diminutive orchids. A few narrow grasslike leaves grow from the base of the plant. Nodding ladies'-tresses orchid, while small, stands out white in the grasses of wetter areas, sometimes by the hundreds, sometimes only a plant or two. A close look-alike, Great Plains ladies'-tresses, sometimes still confuses us, but we've learned that only nodding ladies'-tresses orchid has basal leaves.

One-sided pyrola

Orthilia secunda

We've sometimes mistaken these small (eight-to-ten-inch) plants for orchids. Look closely at the tiny nodding bell-like flowers growing along one side of the stem; although the flowers might look unusual, they don't have the irregular shape of orchid blossoms. The rosette of basal leaves stays green all winter, and the plant is sometimes called sidebells wintergreen.

Ragged fringed orchid

Platanthera lacera

We first saw this plant in bud and were unsure what kind of orchid it was. We visited again a few days later when the pale greenish flowers had opened and knew at once from the deeply fringed lips that it was a ragged fringed orchid. The lower lip of the blossom is divided into three lobes, all of which are deeply divided even more. (We're still searching for purple fringed orchid, which also grows in Minnesota, and we're pretty sure that when we find it, we won't confuse the two.)

Ram's head orchid

Cypripedium arietinum

This small (about eight inches tall), beautiful, rare orchid has a purple-veined lower pouch with a white opening covered with fuzzy hairs. One of the flower's purple sepals arches over the "slipper" like a hood. Minnesota lists it as endangered, and we felt very lucky to see it growing in a northern swamp forest. We'll go back every year to look for this favorite flower.

Small yellow lady's-slipper

Cypripedium parviflorum var. makasin

Yellow lady's-slipper orchids with twisty side sepals come in two sizes, large and small. The sepals of small yellow lady's-slipper are a dark red, while the sepals of large yellow lady's-slipper are striped or speckled. A small yellow lady's-slipper might look large in a photo, but its sunshiny flowers are about the size of a penny.

Spotted Joe-pye weed

Eutrochium maculatum

The spotted part of this plant's name comes from the purple spots that often dot the stem. Large leaves grow in whorls on the stem, and the flowers form clusters of little pale rose-colored blossoms at the ends of the stalks. You might find this towering (up to ten feet tall) plant in wetter parts of the prairie or low, damp, open places in the middle of the woods.

Swamp milkweed

Asclepias incarnata

Minnesota has more than a dozen milkweed species that provide nectar for insects and food for insect larvae, especially monarch butterflies, whose larvae eat only milkweed leaves. Not only butterflies and other insects depend on milkweed—some birds also use old plant stem fibers in their nests. Milkweed plant juice is a milky color, which explains its name. How to tell if the milkweed you are looking at is swamp milkweed: it has clusters of bright reddish-purple flowers, narrow leaves, and prefers damper places.

Tubercled rein orchid

Platanthera flava var. herbiola

We've found these plants growing in the shade of shrubs in a wetland. Yellow flowers crowd the spikes, which rise no more than a foot high. Look closely at the bottom lip of a flower and you'll see a little bump in the center that is called a tubercle. If you find one flowering rein orchid, look around—they tend to grow in patches. They are listed as threatened in Minnesota.

FIELD NOTES
A WORLD AWAY (BY THE HIGHWAY)

Traffic from 35W thrums as we wade through into this protected area on a cool and sunny day. How much wildness can there be so close to the Twin Cities? We want to know, so we slosh on.

Many of the plants are still just promises—when we come back we'll most likely see swamp milkweed and more blazing stars than we've ever seen before in one place, but for now they are simply stalks and leaves and buds. Long, slender stalks of grass beaded with dew bend in the sunlight. Blue flag blooms, and an orange dragonfly flits through the air.

Blaine Scientific and Natural Area is a wetland, but small microhabitats of slightly drier ground rise up like little islands. On one of these we find ragged fringed orchid blooming delicately among wood betony already going to seed. Under nearby shrubby growth, tubercled rein orchids bloom buttery yellow.

Blue flag

We'll come back again to this little bit of wildness so close to a major highway and two cities. Who knows what else we'll find? We've read that nodding ladies'-tresses orchids also grow here in the Blaine Scientific and Natural Area. You just never know what you'll see, even in unlikely places, until you look.

Places to Look

Minnesota Valley National Wildlife Visitor Center in Bloomington has planted prairie by the visitor center building, but hike down the trail to the wetlands and marshes to look for more watery blooms.

Wood Lake Nature Center in Richfield is predominantly marsh with trails and boardwalks, but you might also see woodland flowers or prairie flowers. In early spring we saw the biggest Jack-in-the-pulpit we've ever seen.

Lake Nokomis has several wetland ponds around its edges where, on a walk around the lake, you might see milkweeds, spotted Joe-pye weed, bottle gentian, and blazing star among other native wildflowers and grasses.

Lake Carlos State Park north of Alexandria, located where prairie and mixed northern forest meet, has marshes, ponds, lakes, and even a bog—a great place to look for flowers that don't mind getting their feet wet.

Rice Lake State Park near Owatonna has shallow marshes and lakes among hardwoods and restored prairie. Look for wetland flowers such as marsh marigold, skunk cabbage, swamp milkweed, spotted Joe-pye weed, and gentians. If you visit in the spring, look in the forest for trout lily, Virginia spring beauty, hepaticas, Virginia bluebells, and other early woodland flowers. This is the one park we know of where you can do a drive-by wildflower search.

William O'Brien State Park near Marine on St. Croix has all kinds of wetlands, including tamarack swamp, fens, and marshes. Look for blue flag and other wetland flowers.

Practically all state parks in Minnesota have some kind of wetland to visit, and any of the parks suggested in the 10,000 Lakes section will most likely also have marshy areas to search for native wildflowers.

HIGH SUMMER ON THE PRAIRIE

The Sound and Scent of Summer

Even with your eyes closed, you can still tell a prairie in the middle of summer. Bees buzz, birds call, the scent of blooming milkweed fills the air, a warm breeze blows by. From the earliest pasqueflower to the last goldenrod, something is always in bloom, providing pollen and nectar and larval food for native insects. Small holes in the ground may be nests for native bumblebees. Larger holes might be places where badgers have dug down, tossing rocks around the entrance with their powerful paws. In summer the prairie blossoms with rattlesnake master, wild bergamot, gray-headed coneflower, compass plant, wild quinine, purple prairie clover, several kinds of blazing star and milkweed, big bluestem, black-eyed Susan, and so much more—more than three hundred different species of flowering plants can be found in prairies.

Plants in virgin prairies grow so close and so thickly that invasive plants and trees have a hard time getting a foothold. Although not much virgin prairie is left now, there are still places where you can stand, look up at a vast sky over a rolling hill where no power lines are visible, and imagine yourself back to when prairie ruled the plains.

A SPINY SURPRISE

You might think of cactuses as heat-loving desert plants. We did, until we looked down at our feet in a dry rocky prairie in southern Minnesota and saw cactuses growing there. Three kinds of cactuses are native to the southern half of Minnesota: brittle prickly pear; plains prickly pear, a species of special concern in Minnesota; and ball cactus, a threatened species in Minnesota. Plains prickly pear and brittle prickly pear both have bright yellow blooms; plains prickly pear flowers ripen into pinkish-green fruits, while brittle prickly pear flowers ripen into reddish-brown fruits. Ball cactus (which we have yet to see) has purple flowers and brown fruit.

How do cactuses survive our brutal winters? They grow mainly on rocky outcrops or sandy blowouts that help hold heat in the summer. In the fall, the fleshy cactus plants lose water and shrink into wrinkled pads. This loss of water helps prevent ice from forming in the cactuses, which can damage or kill the plants. Come spring and rain, the plants fill with water again. Cactuses aren't common here, and the ball cactus is the rarest of all, but finding any cactus in a dry prairie or on a rocky ledge is a delightful reminder of Minnesota's plant diversity.

Blooms in Abundance

Black-eyed Susan

Rudbeckia hirta

These bright golden ray flowers with dark disc flowers in the center thrive in sunny places. Individual plants live only one or two years, but they are generous with their seeds and can form large patches of flowers. The leaves and stems are hairy, and the larvae of silvery checkerspot butterflies feed on the leaves.

Blanketflower

Gaillardia aristata

Blanketflower's bright blooms with their yellow- and red-banded petals raying out from a red disc look like a sunburst on the prairie. Blanketflower grows in drier places—you're most likely to see it in the wild in northwestern parts of the state. Because of habitat loss to farming, blanketflower is a Minnesota species of special concern.

Brittle prickly pear

Opuntia fragilis

Brittle prickly pear's pads are smaller, rounder, and break off more easily (one way it spreads) than those of its relative, plains prickly pear. The delicate yellow flowers that bloom later in the summer look almost like tissue paper, and the plant has a fragile (but prickly) feel. Cactus pads are actually modified stems—they store water, produce flowers, and make food through photosynthesis. The spines are modified leaves that protect the plant and also reduce water loss from evaporation.

Butterfly-weed

Asclepias tuberosa

Like other milkweeds, butterfly-weed is a host plant for monarch butterfly larvae, who eat the leaves. Unlike other kinds of milkweed, butterfly-weed has clear sap instead of milky sap and flowers that range from pale yellow-orange to vivid orange to deep red, standouts in the summer prairie grasses. Narrow seedpods that form are filled with small seeds on silky plumes that drift on the wind when the pods split open.

Common milkweed

Asclepias syriaca

Common milkweed has showy clusters of pink flowers and wide oval leaves with soft hairs on the undersides. Flowers ripen into seedpods that open to release seeds on silky parachutes that ride the prairie wind. Like other milkweeds, this is a plant that monarchs lay their eggs on and that their larvae eat until they are ready to metamorphose into butterflies. If you find a large green-and-black-striped caterpillar on a milkweed, listen closely. You might hear it chomping its way through a leaf on its way to becoming a butterfly.

Compass plant

Silphium laciniatum

Compass plant's name comes from its large (up to eighteen inches), deeply lobed leaves, which orient themselves north and south so that the hot, drying sun doesn't shine directly on the leaf surfaces. The plants can grow up to ten feet tall, and the five-inch-wide yellow ray flowers around a central yellow disc grow at the top of the stem and in the leaf axils. In full bloom this is a stunner on the summer prairie, but its distinctive leaves make it recognizable any time of year.

Culver's root

Veronicastrum virginicum

Finding Culver's root growing in the prairie always reminds us of an escapee from a Dr. Seuss book with its crazily curving tapered spires of little white flowers on tall stems. Flowers on the spike bloom from the bottom up. Look for it in wetter places in the prairie, and when you find it, look closely for bees busy among the blossoms.

Flowering spurge

Euphorbia corollata

The flower stalks of flowering spurge radiate out above circles of five leaves so that the flowers form another circle of small white blossoms and make us think of little umbrellas. Flowering spurge can grow up to three feet tall, but its leaves and flowers give the plant a delicate look.

Gray-headed coneflower

Ratibida pinnata

Long, narrow yellow petals (which are actually ray flowers) hang down around a gray-green central cone that is actually a disc of flowers and that eventually turns brown. Each composite flower grows on its own stalk, but a single plant might have up to a dozen blooms. Stems are hairy, and the rough-feeling leaves are divided into long lobes.

Indian paintbrush

Castilleja coccinea

Flowers at the top of Indian paintbrush are in a dense spike, but what we actually see and often think are the flowers are the yellow, orange, or red bracts that grow under and around the flowers. The yellow-green flowers themselves are less than an inch long and almost hidden by the colorful bracts. This is a partially parasitic plant, getting some of its nutrients through its roots from other plants, usually prairie grasses.

Leadplant

Amorpha canescens

This woody-stemmed plant stands out on the prairie, not just for its size (up to three feet tall) and gray-green hairy compound leaves but also for its dense, long, slender clusters of purple blossoms, each with a single petal, with bright orange-tipped stamens. Leadplant's roots are long and tough, and settlers trying to plow the prairie sometimes called leadplant roots prairie shoestrings or Devil's shoestrings. Leadplant is an indicator of high-quality prairie. Its name comes from the belief that where it grows, you can find lead in the ground.

Michigan lily

Lilium michiganense

Michigan lily has up to eleven flowers, each on its own stalk. The long, narrow leaves grow in whorls on the stem and the flowers hang down, while the tepals (petallike parts) curve gracefully upward. This is one of the few truly orange prairie flowers in Minnesota.

Narrow-leaved purple coneflower

Echinacea angustifolia

Other coneflowers grow in Minnesota, but narrow-leaved purple coneflower is the only purple coneflower native to the state. Pinkish-purple ray flowers surround a spiny center made up of disc flowers. Part of the scientific name, *Echinacea*, comes from the Greek word *echinos* for hedgehog (a spiny mammal) because of this spiny central disc.

Pale-spike lobelia

Lobelia spicata

The tiny delicate blue or white flowers of pale-spike lobelia bloom along a single stem up to a foot or more tall. The flowers themselves are less than half an inch across; look closely and you'll see that each flower has a small upper lip divided into two pointed lobes and a large lower lip divided into three pointed lobes. You might find pale-spike lobelia blooming in the prairie grasses anytime from spring until late summer.

Plains prickly pear

Opuntia humifusa var. humifusa

Minnesota has two kinds of prickly pear cactus plants, plains prickly pear and brittle prickly pear. Both survive winters by losing water from their pads. The shriveled pads (larger on plains prickly pear) swell again in spring. Plains prickly pear has lovely yellow flowers that ripen into pinkish-green waxy fruits. Look for them in dry prairies and sandy places in southern parts of the state. Plains prickly pear is a species of special concern in Minnesota.

Prairie blazing star

Liatris pycnostachya

While many prairie plants bloom from the bottom up, the tall spikes of prairie blazing star, like other blazing stars, bloom from the top down. Lower on the stem, the plant's narrow leaves are longer, becoming smaller and smaller higher up the stem. Although we've seen a few white flowering plants, blazing star flowers are almost always some shade of rose to purple.

Prairie coreopsis

Coreopsis palmata

Prairie coreopsis has bright yellow ray flowers around a yellow center made up of disc flowers. We love one of this plant's common names, bird's foot coreopsis, because the deeply lobed leaves remind us of bird tracks. One explanation for yet another common name, stiff tickseed, is that the seeds resemble ticks and stick tight like ticks.

Prairie ragwort

Packera plattensis

Clusters of yellow flowers grow at the top of the stems of prairie ragwort plants. What look like petals are actually ray flowers surrounding the center, which is made up of small disc flowers. The plant's stalks and leaves are covered with cobwebby white hairs as though spiders have been hard at work.

Prairie rose

Rosa arkansana

Prairie rose bears lovely white-to-pink five-petaled flowers with yellow centers, but you can spot it even on the fall prairie by the red rose hips that form after flowering. Prairie rose plants are the shortest of Minnesota's four native roses, growing up to three feet tall, with bristly stems and leaves made up of several jagged-edged leaflets.

Prairie sunflower

Helianthus petiolaris

Prairie sunflower can grow up to several feet high, but the flowers are usually no more than three inches across. We've long had a default category for sunflowers ("Oh, that's some kind of helianthus"), but we can tell prairie sunflower by its smaller leaves, which have only a few teeth and which alternate on the stem.

Purple prairie clover

Dalea purpurea

Each tiny quarter-inch flower of purple prairie clover has four even tinier petals attached, and the orange stamens of the flowers extend out past the petals. Flowers open from bottom to top so that when the flower spike first starts to bloom it looks to be wearing a tutu. While the plant can grow up to three feet tall, its taproot may grow as deep as six feet down into the soil.

Rattlesnake master

Eryngium yuccifolium

Small prickly globes covered with tiny white flowers bloom on stalks that radiate from the stem of rattlesnake master. The long, stiff, pointed leaves are edged with little spikes. There's no mistaking these distinctive plants, even when the globes turn brown with seed in the fall. According to one field guide, shoes woven from leaf fibers of rattlesnake master have been found in a cave and dated at 8,300 years old. Rattlesnake master is a species of special concern in Minnesota.

Rough blazing star

Liatris aspera

Purple flowers along the single stems of rough blazing star look almost like little pom-poms as they bloom from the top to the bottom of the stem. Up to four feet tall, these plants stand out on the prairie, especially in the fall when migrating monarchs flock to them—we've seen seven monarchs at a time on one blooming plant.

Round-headed bush clover

Lespedeza capitata

Here's a plant that hard to misidentify. True to its name, the tiny white flowers grow in dense round, mostly green heads. Even when the heads have turned brown, you can pick out this tall distinctive plant on the autumn prairie. Compound leaves have three leaflets and might remind you of the clover that often grows in lawns.

Sullivant's milkweed

Asclepias sullivantii

Sullivant's milkweed has clusters of pink flowers with five downward-pointing petals and five pink hoods over them. Its oval-shaped leaves are smooth and feel slightly waxy, with a red vein down the center of the underside of the leaf. This rare milkweed is an indication of less-disturbed prairie.

Western prairie fringed orchid

Platanthera praeclara

Minnesota has the largest population in the country of this federally threatened and state endangered species. Like most orchids, the white flower has an arching hood. Its lower lip is divided into three fringed lobes, although not as divided as that of ragged fringed orchid. Finding this delicately beautiful bloom on the summer prairie is a special gift.

White camas

Zigadenas elegans

The narrow grasslike leaves of white camas can be up to a foot long, but it's the six-petaled white flowers that catch the eye. The flower's petals have small greenish-yellow glands that give the inside of the flower a star-shaped look. The plant gets one of its other common names, death camas, because every part of the plant is poisonous.

White prairie clover

Dalea candida

Like purple prairie clover, the densely packed flowers on the cylindrical flower spike of white prairie clover bloom from the bottom up, like a wreath of white moving upward. Although you might see both purple and white prairie clovers in the same habitat or same prairie, white prairie clover begins blooming earlier and is less common. White or purple, we love finding these flowers in their dancing skirts.

White wild indigo

Baptisia lactea

White wild indigo has tall, slender spires with showy white flowers scattered along them and leaves made up of three leaflets each. The plant keeps adding flower buds to the top of the spike as it blooms, so you might see white wild indigo blooming for a long time in the summer prairie. Bumblebees pry open the flowers and work their way from flower to flower up the stem. This is a species of special concern in Minnesota.

Wild bergamot

Monarda fistulosa

Wild bergamot is a mint and has that plant family's distinctive square stem and irregularly shaped flowers. Like other mints, wild bergamot tends to spread, and you may see it growing in crowded clumps of blue-purple tufted tiny tubular flowers. This is one plant you might recognize even after the flowers fall away—the center is a round brown ball of calyxes that contains the seeds, and the leaves are still fragrant.

Wild quinine

Parthenium integrifolium

Wild quinine flummoxed us at first—the cluster of small, dense-looking individual flower heads look a little like pentagons with wings. Each flower head is made up of five or six ray flowers surrounding the disc flowers in the center. Minnesota, where it is listed as endangered, is the northernmost edge of wild quinine's range. You might find it in undisturbed (unplowed and ungrazed) prairies or along railroad rights-of-way—or even in a lucky gardener's yard.

Wood lily

Lilium philadelphicum

Don't be misled by the common name—you might find this flower in wooded areas, but this is mainly a prairie flower. Whereas Michigan lilies open pointing downward, the tepals (petallike parts) of wood lily make a cup open to the sky. Spaces between the petals and sepals let rainwater drain out of the cup. Monarch butterflies, swallowtail butterflies, great spangled fritillary butterflies, and hummingbirds all visit these vivid orange flowers with dark spots.

Yarrow

Achillea millefolium

Scientists disagree on whether yarrow is native or alien; some list a shorter native version and a taller version introduced from Europe. Native or introduced, you can find yarrow in high-quality prairies as well as weedy lots, and you might see native bees visiting its flowers. Tiny white-to-pink flowers form tight clusters at the tops of stems, and yarrow's much-divided curly leaves look ferny; part of its scientific name—*millefolium*—means many-leaved.

FIELD NOTES IN SEARCH OF THE WESTERN PRAIRIE FRINGED ORCHID

August

I'd seen a single western prairie fringed orchid once before on a day when the hot sun pressed down and damp heat rose from the ground like a hand smack, in a prairie where no western prairie fringed orchid had been seen for twenty-nine years. When Kelly and I returned to photograph it, I was sure I could find the orchid again. Wasn't it three feet tall and glowing? But the orchid was done blooming, and grasses hid it completely. It took several years of looking before we found that single orchid again, crouched in the grass and gone to seed. We had a yen to see more of these orchids (endangered in Minnesota and threatened nationally) in full bloom, so in August 2016 we set off on a road trip to Minnesota's northwestern prairies, Kelly with her camera and me with directions from a naturalist friend who had seen them there.

We passed two Scientific and Natural Areas on our way north, and we stopped to see white camas, small white lady's-slipper, wood lily, pale-spike lobelia, blazing star, and Indian grass.In high summer and with high hopes we drove on.

"You can see them from the road," the naturalist friend had told us, and while we don't recommend drive-by wildflower searches, when we reached the wildlife area where the orchids were supposed to be, we drove slowly up and down the roads looking for anything white. We saw plenty of white flowers but no orchids.

Up and down, up and down the deserted roads we drove. Just as we were about to give up, Kelly pointed down into the roadside ditch.

"What's that?" she asked.

That was more than a dozen western prairie fringed orchids in full and radiant bloom. We were radiant, too, at finally finding what we'd come to see. Kelly took picture after picture as we lingered, unwilling to leave these rare orchids, well worth the miles we'd traveled to see and more beautiful than we had imagined.

Places to Look

Fort Snelling State Park, where the Minnesota and Mississippi rivers converge, has a variety of habitats with a variety of flowers. In summer prairie, look for butterfly-weed, prairie rose, wild bergamot, blue vervain, wood lily, milkweeds, and pale purple coneflower.

Hyland Lake Park Reserve in Bloomington has prairie where wild bergamot, goldenrods, and other wildflowers grow.

Crow-Hassan Park Reserve by Rogers has six hundred acres of restored prairie that offer all sorts of prairie flowers, including butterfly-weed, blazing star, several kinds of goldenrod, wild bergamot, leadplant, long-bracted spider-wort, prairie coreopsis, prairie grasses, and coneflowers.

Big Stone Lake State Park not far from Ortonville includes Bonanza Prairie, a Scientific and Natural Area, within its boundaries. Look for all sorts of prairie flowers, including compass plant, pale purple coneflower, prairie rose, white prairie clover, purple prairie clover, butterfly-weed, wild bergamot, and prairie blazing star. The prairie overlooks Big Stone Lake, the source of the Minnesota River.

Blue Mounds State Park near Luverne is Minnesota's only state park where you can find both kinds of prickly pear cactus right along the trail. You might also see blue vervain, narrow-leaved purple coneflower, prairie sunflower, compass plant, purple prairie clover, wild rose, leadplant, round-headed bush clover—and a herd of bison.

Minneopa State Park near Mankato is partly prairie, where you can look for black-eyed Susan, butterfly-weed, common milkweed, gray-headed coneflower, leadplant, pale purple coneflower, purple prairie clover, rattlesnake master, rough blazing star, and white wild indigo. The park also has two waterfalls and a bison herd. Check with the park to confirm which days you can drive through the bison range.

Myre-Big Island State Park near Albert Lea offers wetlands, marshes, prairie, and forest. You might come here for the spring wildflowers and return for the prairie in bloom with big and little bluestem grasses, Indian grass, leadplant, rattlesnake master, prairie clovers, prairie smoke, blazing star, black-eyed Susan, and coneflowers.

You can also visit any of the places listed under Early Prairie or Prairie Fall to see high summer prairie bursting with blooms.

SUMMER

10,000 LAKES (MORE OR LESS)

Up to the Lake

Minnesota lakes are magical places. In *Listening Point,* writer and conservationist Sigurd Olson wrote, "Water reflects not only clouds and trees and cliffs, but all the infinite variations of mind and spirit we bring to it." We owe our thousands of lakes (anywhere from ten thousand to fourteen thousand, depending on how big you define a lake to be) in large part to glaciers, huge sheets of ice that crept across the landscape thousands of years ago and then melted away.

Those glaciers laid the groundwork for Minnesota's many lakes. Glaciers gouged out rocky depressions that filled with water. At its edges a retreating glacier often dropped huge deposits of soil and sediment, burying blocks of ice that later melted, leaving holes called kettles to fill with water and become kettle lakes. Most of our lakes are ten thousand to twelve thousand years old, the end of our last glacial age.

Thanks to the glaciers, more than eleven thousand lakes, from ten acres to 288,800 acres in size, dot the landscape. Fish swim, otters splash, birds nest and dive in them. And people, when they are able, often go "up north" to the lake to camp, swim, stay in cabins, canoe, boat, fish, and be outside.

Even on lakes where cabins crowd the beaches, you might still find native flowers growing, some along the lake edges, some floating, some anchored on the lake bottom.

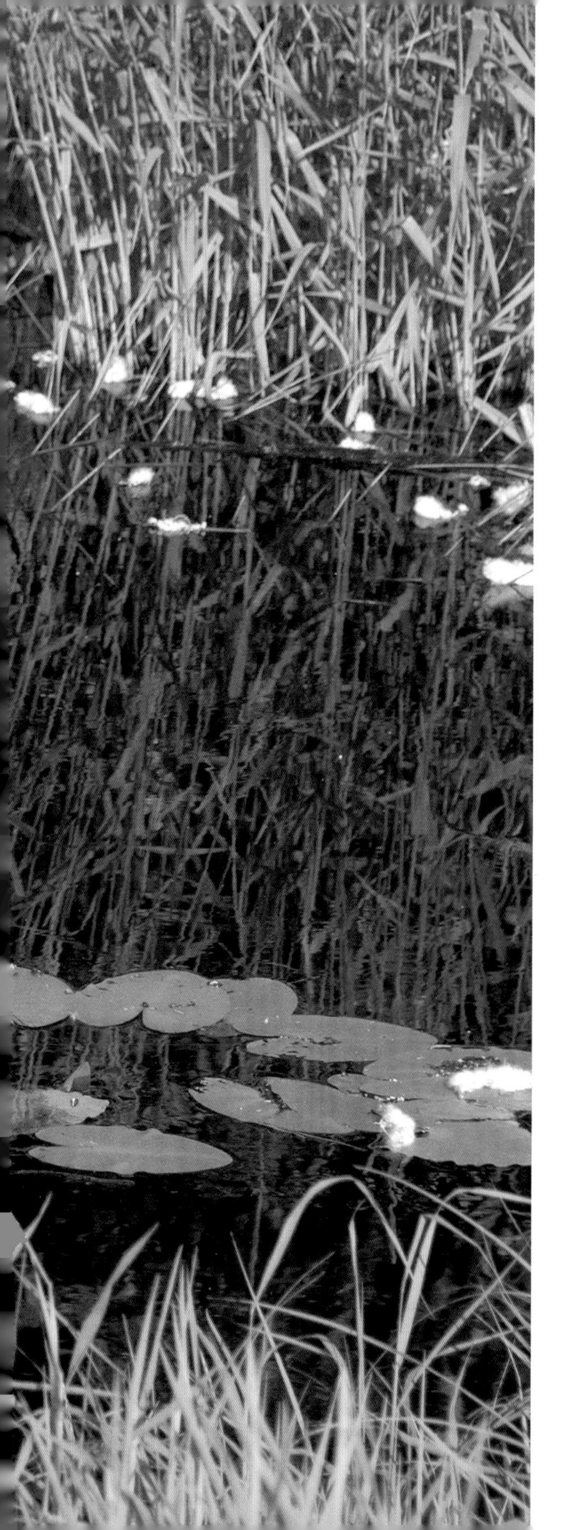

A Temporary Landscape

You might think of lakes as permanent features of the landscape, but most lakes are on their way to becoming land, even though the process might take hundreds or thousands of years. Lakes slowly fill up with sediment and plants, and the edges of a lake may gradually change into a marsh or a fen and eventually to more solid ground. You might not see a lake filling in any time soon, not even in your lifetime, but lakes, in geological time, are temporary features.

The flowers you might see depend, too, on how deep the water is where they grow.

Close to shore you might see plants associated with bogs, such as sundews or buckbean. Some lakes even have floating bogs around their edges, building up peat that will eventually (over a long, long time) become land where shrubs and even trees will grow.

A little farther out in shallow water you might see flowers that have their roots anchored to the bottom such as water lilies or smartweeds. Certain plants don't need a place to put their roots—common bladderwort, for example, has tendrils that hang down in the water but no roots to anchor the plant.

Underwater, too, plants grow, but we've not gone looking for them yet. Whether above the surface or below, our lakes are so much more than simply water.

Flowers Afloat (and Some with Roots)

American white waterlily

Nymphaea odorata

American white waterlily leaves float on the surface of lakes, ponds, and rivers, and so do their elegant and glorious flowers, which are up to six inches wide. When water levels are low, you might see the flowers sticking up several inches above the water. Sometimes American white waterlilies grow so thickly that their large leaves seem to cover the lake and make for challenging paddling. The flowers open in the morning and close at night. Part of their scientific name, *odorata*, is Latin for fragrant—which they are.

Broad-leaved arrowhead

Sagittaria latifolia

The pointed leaves of this plant that stands tall in shallow water give broad-leaved arrowhead its common name. Three-petaled white flowers bloom along the stem. Broad-leaved arrowhead is also called swamp potato because geese and other waterfowl eat its starchy roots and muskrats store them for winter food.

Common bladderwort

Utricularia vulgaris subsp. macrorhiza

Although common bladderwort's bright yellow two-lipped flowers are small, they are easy to spot from a canoe or boat because their stems lift them up above the water. The plant has no roots. This is one of Minnesota's carnivorous species—underwater bladders suck in and digest tiny aquatic organisms. The bladders also fill with air to keep the plant afloat, then fill with water and sink the plant during the winter.

Marsh skullcap

Scutellaria galericulata

Look closely at marsh skull-cap's pairs of small blue tube-shaped flowers rising from each leaf axil and you'll see that the lower lip is purple with a white center speckled with blue dots. These plants belong to the mint family, almost all of whose members have square stems.

Water smartweed

Persicaria amphibia

The small, crowded flower spikes of water smartweed (sometimes called swamp smartweed) stick up above the water like bright pink thumbs while their oval leaves float on the water's surface and the stem stays underneath the water. You might find water smartweed growing on land as well. In its land version, water smartweed has an upright stem and narrow leaves with wavy edges and pointed tips. Water smartweed won't make you smarter, but seeing its vivid blossoms always brightens our day.

Wild calla

Calla palustris

Another name for this elegant plant is water arum. Like other members of the arum family, such as Jack-in-the-pulpit, wild calla has a spathe or large leafy bract. Wild calla's broad white pointed spathe makes the plant a standout in watery places. The smaller central part of the flower (spadix) is packed with tiny blossoms.

Yellow pond-lily

Nuphar variegata

The round flowers of yellow pond-lily may be small, only about three inches across, but they are vividly yellow. Like American white waterlily, the leaves float on water and are notched where the stem joins the leaf, but unlike American white waterlily, yellow pond-lily flowers always stick out of the water several inches and never open completely. The distinctive flower center has been described as looking like a toadstool, and we agree. Along with the tubers of other aquatic plants, yellow pond-lily tubers are food for muskrats.

FIELD NOTES
SUMMERTIME SURPRISE

I've stopped at Kelly's lake cabin overnight on my way home from a trip up north. Cabins line half of the lakeshore, and boats putter back and forth. Sun, water, trees, a pier, people paddleboarding and sunbathing—a typical summer day at many of Minnesota's lakes.

We bask in the sun, then wander through a nearby woods to a small bog. Toward evening, we paddle the canoe to the south end of the lake where a narrow channel leads into a backwater, the kind of place where we love to poke around. Among lily pads and reeds we find the bright pink of water smartweed and, a few inches above the water, small yellow flowers on stems—common bladderwort, one of Minnesota's four kinds of carnivorous plants. Even on a lake where folks fish and boat and swim, Minnesota's tiny hidden treasures grow.

The sun is sinking. The mosquitoes are rising. We paddle back to the beach to watch a blue moon, the second full moon of the month, float up into the evening sky.

Places to Look

So many lakes in Minnesota—how do you choose? Almost any lake offers the chance to look for watery wildflowers, but here are a few to get you started.

Cedar Lake in Minneapolis connects to Lake of the Isles, which connects to Lake Calhoun. (There's a reason Minneapolis is called "city of lakes.") Around these lakes you'll see water plants such as American white waterlily and blue flag as well as prairie plants such as butterfly-weed. And what's not to love about walking around a lake?

Crosby Farm Regional Park (the largest natural park in St. Paul) hugs the Mississippi River and has woodlands, marshes, Crosby Lake, and forests. Among the plants you might see are spotted Joe-pye weed, water lilies, blue flag, and water smartweed. When we visited, the boardwalk was partly underwater, so we waded through tiny tadpoles swimming around our feet. Next time we'll remember to bring boots.

Lake Maria State Park near Monticello has a wealth of marshes and lakes where turtles and swans swim. Lake Maria is part of the Big Woods biome, so it's also a great place to visit for spring woodland flowers and fall colors.

Bearhead State Park near Ely is almost like a trip to the Boundary Waters Canoe Area Wilderness (which is close by). Hike the lakeside trails under the pines or paddle a canoe and look for flowers afloat.

Glendalough State Park near Battle Lake has five lakes and is threaded by trails so you can wildflower search from the shore or the water. Because the park is located where prairie and woodland meet, you might find a wide variety of wildflowers.

SUMMER

MINNESOTA PEATLANDS

Bogs Big and Small

Bogs are some of our state's wildest places, vast stretches of moss and bog plants with trees growing on raised islands within the bog. Northern Minnesota contains the largest stretch of peatlands in the United States outside of Alaska. In plains flattened by glaciers and in scattered basins with poor drainage, sphagnum mosses grow. As the moss dies, it sinks and forms beds of peat moss up to twenty feet or more deep. Water in bogs is usually cold and acidic, and few plants can survive there. The plants that do grow in bogs have adapted to the harsh conditions. Floating bogs also develop along the edges of some lakes, gradually forming mats of peat where bog flowers bloom.

For all their harshness, bogs are actually delicate wildernesses. A single footprint in a bog can last a long time. In Big Bog State Recreation Area, you can still see trails left by caribou almost a hundred years ago and the ditches that were dug in the early 1900s in a mistaken attempt to drain the bog and turn it into farmland. If you go to a bog, look closely and go quietly. Let the bog's strange wildness settle over you. Life-and-death struggles can go on even in the pitcher of a purple pitcher plant, which traps unwary insects and dissolves them in its depths.

Purple pitcher plant

Hungry Plants

Minnesota has four main kinds of carnivorous plants that grow in bogs and other habitats—purple pitcher plant, butterwort, sundews, and bladderworts. These plants trap insects and digest them for nitrogen; they also make their own food from chlorophyll like other green plants.

Purple pitcher plants, found in bogs, are Minnesota's largest carnivorous plants. Bright colors and nectar secretions lure insects down into the plant's pitcher (actually a modified leaf) where downward-pointing hairs prevent the insects from crawling out again. Eventually, the insects fall into the pool at the bottom of the pitcher and are slowly digested. Not all insects are devoured by purple pitcher plants—for some insects, purple pitcher plants are part of their life cycle. The larvae of one species of mosquito can live only in the pitcher plant's pitcher, unharmed by the enzymes that digest other prey. Some spiders spin webs above purple pitcher plants to catch the insects lured to the plant before they can fly or fall inside. Do the spiders know they'll catch more insects this way? We don't know, but we've seen those webs spun right over some purple pitcher plant pitchers.

Sundews, found in bogs and along the edges of some lakes, trap insects in sticky drops on the ends of stalks or tentacles that dot the plants and glisten red in the slanting sunlight. (Part of the plant's scientific name, *drosera,* is from a Greek word meaning "dewy.") The longer stalked glands on the edges of the leaves press the stuck insects down onto the leaf itself, where they are digested. Interestingly, some studies have found that sundew tentacles aren't triggered by inorganic matter, such as sand, just by insects.

Although they are not bog plants, bladderworts and butterwort are also carnivorous.

Bladderworts, which float rootlessly in the shallow waters of lakes, have tiny underwater bladders. Hairs on a bladder are triggered by the touch of even tinier insects, which causes the bladder to inflate, sucking in water along with the insect. The plant expels the water and slowly digests the insect while it waits for its next unwary prey.

Butterwort, found in rocky cracks and crevices along Lake Superior's North Shore, has shiny-looking leaves covered with sticky droplets. When an unsuspecting insect lands on the leaf, it sticks fast. The leaf slowly curls around the insect and absorbs it. One of Minnesota's arctic relicts, this is the rarest insect-eating plant in the state.

Why do these plants eat insects? One theory is that carnivorous plants get needed nitrogen and phosphorous from insects—a kind of vitamin pill for plants. We do know this: if you are not an insect, it's safe to go looking for these carnivorous plants.

Plants in the Peat Moss

Bog laurel

Kalmia polifolia

These brilliant pink cuplike flowers often grow in clusters among the mosses of a peat bog. The shiny leathery-feeling leaves curl under at the edges, and the petals look almost as though they've been stitched together. The tips of the flower's stamens are embedded in the surface of the petals, a kind of spring-loading so that when an insect lands on the flower, the stamen tips are released and pollen is hurled onto the insect.

Bog rosemary

Andromeda polifolia

Clusters of small pink flowers hang like bells on bog rosemary, and the leaves stay on all winter, which helps the plant save the energy it would take to grow new leaves each spring. The narrow leaves curl under at the edges, and although they might resemble the garden herb rosemary, the plants are not related. True to its name, bog rosemary grows only in a bog, where the plant has adapted to the cold, acidic environment.

Buckbean

Menyanthes trifoliata

Buckbean flowers bloom along a spike at the top of a tall stalk. Slender, hairy, starkly white petals give the flowers a lacy look. Wide leaflets in groups of three often grow on separate stalks a few inches away from the flower stalk. This is a striking plant to spy, often growing in colonies of white. We've seen it in bogs, but fens and lake edges are also buckbean habitat.

Dragon's-mouth

Arethusa bulbosa

Dragon's-mouth is another unmistakable orchid. Bright pink lateral petals and sepals curl out to form a hood over the orchid's large lower lip, which is frilly white with purple markings and a white or yellow "beard." This brilliant little orchid likes open peatland, including floating bogs. Bemidji State Park's bog boardwalk is the only place we've ever been lucky enough to see it.

Fairy-slipper orchid

Calypso bulbosa var. americana

Look for this exquisite tiny orchid, also called calypso, in northern cedar swamps and forested bogs in late May and early June. Although this orchid is small (three to six inches tall), its vivid pink-to-lavender colors help make it visible on the mossy ground. On the rare occasions when we've found one—one amazing day we saw sixteen blooming—we marvel at their loveliness.

Labrador tea

Rhododendrom groenlandicum

The flowers of Labrador tea grow in spidery white clusters. The edges of the leaves roll under, and the undersides of new leaves are wooly white while the undersides of older leaves look rusty brown. This plant really can be made into a tea, but don't try it—besides the fact that you should leave native plants unharmed, the tea is said to be toxic.

Leatherleaf

Chamaedaphne calyculata

True to its name, the leaves of this bog shrub have a leathery look and texture and many—but not all—leaves last through a second summer. Leaves that fall off the plant and decompose help to turn bog water brown with their tannins. Tiny bell-like tubular flowers bloom early and hang in an orderly row from a stalk.

Purple pitcher plant

Sarracenia purpurea

The single large flower on a purple pitcher plant looks like a pinwheel toy. The pitchers are actually brightly colored pitcher-shaped leaves that lure insects in where slippery slopes and downward-pointing hairs prevent them from crawling out. When the insects drown in the liquid at the bottom of the pitcher, the plant digests them. Because pitcher plants are often sunken in the moss, we have to train our eyes to find them—and once we've seen one, we usually discover many more.

Round-leaved sundew

Drosera rotundifolia

We looked for a long time for round-leaved sundew thinking, based on photos in guidebooks, that this distinctive carnivorous plant would be easy to recognize. When one day we finally saw it along a floating bog, we realized how very, very tiny these tentacle-leaved plants are. We haven't seen round-leaved sundew flowering yet, but guidebooks show one white flower at a time blooming from the bottom to the top of the flower stalk. Once you spot your first sundew and know what to look for, you'll find yourself seeing more of them in lake edges and bogs.

Small cranberry

Vaccinium oxycoccos

This shrubby plant creeps along the mossy surface of peatlands. Its pink flowers with their folded-back petals look similar to jeweled shooting stars and ripen into red berries. Bog copper butterflies lay their eggs on these plants, and their larvae feed on them.

Three-leaved false Solomon's seal

Maianthemum trifolium

When we first saw three-leaved false Solomon's seal's lacy flower spike, we wondered why Canada mayflower, a woodland flower, was growing so thickly in a bog. On closer look, we saw that the long leaves hugged the stem and the small white flowers in the cluster at the top of the stem had six tepals, not four like Canada mayflower. A delicate surprise in the mossy bog.

Tuberous grass-pink

Calopogon tuberosus

We often look at orchid leaves for help in identifying the plants, and tuberous grass-pink orchid has a single long, wide grasslike leaf, but the flower's vivid pink triangular-looking petals also make it easy to recognize. The flower of tuberous grass-pink has a wide-lipped upper petal rather than a wide lower lip like other orchids, making it appear upside down—even though the flower is actually right side up.

Tussock cottongrass

Eriophorum vaginatum

Cottongrass is really a sedge, not a grass. Sedges differ from grasses, and a general rule (but with exceptions) for telling whether a plant is a sedge or a grass is that "sedges have edges while grasses have joints." Run your fingers along the stem of tussock cottongrass and you'll feel the three distinct edges of its triangular-shaped stem. Tussock cottongrass has a single spikelet or tuft at the top of the stem. The wooly white heads that look like a field of summertime snow are actually seed heads, not flowers.

FIELD NOTES FALLING IN LOVE WITH THE BOG

September

I've contracted to do a book about Minnesota habitats, and while I've spent time in many of Minnesota's outdoor places and read about others, no research can really tell how a place feels or smells or looks in all its tiny details. So, we head out for the Big Bog in northwestern Minnesota, part of more than five hundred square miles of peatland and the largest bog in the lower forty-eight. I don't expect to see much—just wide, flat areas with lots of mosquitoes.

Bogs are made of sphagnum moss, sometimes twenty feet thick, which keeps the water acidic and cold. Walking on moss is difficult and tough on the moss as well. Thankfully, the Big Bog has a boardwalk a mile long, so we follow it into the bog, which is surprisingly mosquito-free.

Before we even reach the end of the boardwalk, where shades of red and green moss stretch as far as we can see with little islands of trees like ships sailing along and purple pitcher plant leaves shining in the sunlight, I have fallen in love with the bog. Who knew plants could grow in such tough conditions? Who knew that the needles of a tamarack tree were so soft? Who knew that the silence over a bog is like nothing we've ever heard before?

On the walk back, I am already scheming a picture book on the bog so everyone can know about this amazing place.

Before we leave, we follow another path through the woods along a marshy area at the edge of the bog. And here we find the mosquitoes, so many we inhale them with every breath. We turn and run for the car.

A Bog Close to Home: Spring

Most of Minnesota's bogs and peatlands are in the more northern parts of the state, but a twenty-minute drive from our houses usually brings us to Eloise Butler Wildflower Garden with its quaking bog, a small but rich bog remnant. Today, though, highway construction and closed-off roads send us on a tortuous path—even close to home we can still find a way to get lost.

Persistence pays, however—when we finally arrive, the bog is luminous in the early-morning light, mosses so deeply green we want to sink into them. We stay on the boardwalk, however, because careless footprints do so much damage to a bog.

A few weeks ago, we came here to find and photograph leatherleaf blooming; now the plant has only a few spent blossoms left on it. Today we're in search of bog rosemary, an evergreen shrubby plant with clusters of small flowers. Light slants through the pine trees, and the tamaracks are getting their soft, bright new needles. Buckbean blooms in white spires, and the leaves of wild calla grow in great bunches with only a few flowers beginning to bloom. Labrador tea flowers are in bud, tiny white violets are scattered in the moss, and the leaves of starflowers are everywhere, although only one small, white star-shaped flower has opened.

Purple pitcher plant

We don't find bog rosemary—not stem or leaf or blossom or bud. But we don't regret this trip. We'll come back again to see the wild callas and starflowers abundantly blooming. Today it's enough to walk through the green falling light, listen to the gargling call of a bird, and see the richly colored mosses and the promise of flowers soon to bloom.

We love having this bog almost in our own backyard.

Places to Look

Eloise Butler Bog in Theodore Wirth Park in Minneapolis by Golden Valley is a small gem of a bog boardwalk where we have seen leatherleaf, wild calla, and buckbean aplenty. Look for marsh marigold, broad-leaved arrowhead, Labrador tea, and starflower. Even close to a busy urban freeway the bog seems to soak up sound and wrap us in silence.

Big Bog State Recreation Area has a mile-long environmentally sensitive boardwalk from which you can see purple pitcher plant, small cranberry, tussock cottongrass, bog laurel, and leather-leaf. Orchids and sundews are said to grow there, too, but we've yet to spot them. At the end of the boardwalk you may feel as though you are in the wildest place you have ever been. We did.

Lake Bemidji State Park has an amazing boardwalk where, depending on the time of year, you might spot Labrador tea, round-leaved sundew, purple pitcher plant, showy lady's-slipper, stemless lady's-slipper, bog rosemary (the only place we've ever seen it blooming), tuberous grass-pink, dragon's-mouth, and buckbean. A thermometer display shows the striking difference between air temperature and the temperature ten inches below the surface of the bog. We were there on a 73-degree day, and the bog below-surface temperature was 38 degrees.

Mille Lacs Kathio State Park has a boardwalk that leads through a tamarack bog where you can look for bog laurel, Labrador tea, leatherleaf, stemless lady's-slipper, tussock cottongrass, three-leaved false Solomon's seal, and other bog plants. On our way to the bog when we visited in the spring, we passed more trilliums than we had ever seen.

Savanna Portage State Park near McGregor has a boardwalk through forested and mossy bog to overlook Bog Lake. From the overlook we spotted Labrador tea, purple pitcher plant, tussock cottongrass, bog laurel, buckbean, and leatherleaf. The trail to the boardwalk skirts a lake that made us think we were in the Boundary Waters.

AUTUMN

AUTUMN WOODS

Sunlight on the Forest Floor

In autumn, most deciduous (leaf-bearing) trees flame into beautiful colors before their leaves fall to the forest floor. Some of these colors have been hidden inside the leaves, masked by the green chlorophyll that makes food from sunlight, air, water, and nutrients. As days shorten and nights grow longer, the cells where the leaves join their stems thicken and block off the flow between leaves and tree, causing the chlorophyll to break down. When the green chlorophyll disappears, other colors emerge—yellows and oranges that have been hiding all along and reds and purples made from sugar trapped in the leaves. As these colors, too, break down, only the brown tannins are left. Eventually, the leaves fall to the forest floor, where they will become new soil.

ANOTHER KIND OF COLOR

Minnesota's fall colors are caught in the leaves of the trees, but don't forget to look down around your feet. Even though spring is the time to search for woodland wildflowers, in late summer and early fall you might see white rattlesnake-root blooming in shaded places, and later still large-leaved aster and downy rattlesnake-plantain bloom as the forest winds down toward the long winter.

And although spring's flowers are long gone, some of the plants we've seen earlier, the ones that haven't ephemerally vanished, still offer up color. Jack-in-the-pulpit's flower has ripened into a cluster of bright red berries. Bunchberry now has a bunch of red berries, and bluebead lily's yellow flowers have become blue bead-like berries. Red berries hang under the leaves of giant Solomon's seal and droop from the stalks of false Solomon's seal. Rose twisted-stalk's seedy berries are purplish red, and nodding trillium now has a nodding rosy berry about an inch long. Poison ivy may look lovely with its red leaves and later in the season its clusters of cream-colored berries, but beware—you can still get a rash. Hepatica leaves that persist through the winter are already mottling to purple. You might need to look more closely for these colors of fall, but unless the birds and other animals have beaten you to the berries, you'll still find color on the forest floor.

Jack-in-the-pulpit

Poison ivy

Bunchberry

Forest Flowers in Fall

Downy rattlesnake-plantain

Goodyera pubescens

You can recognize downy rattlesnake-plantain orchids by the distinctive silver lines and patterns on their clusters of basal leaves. Because these orchids spread by rhizomes as well as seeds, rattlesnake-plantain can form blankets of plants. The orchid spikes are small and open later in the summer. Even if you don't catch this plant flowering (and we've tried . . . and tried . . . and tried), the leaves will tell you its name.

Large-leaved aster

Eurybia macrophylla

Most woodland flowers are spring bloomers, but large-leaved aster blooms from late summer into fall in forest openings. You can spot large-leaved aster throughout the summer before it blooms by its giant-sized basal leaves up to seven inches long and five inches wide. The flowers, when they open, have ten to twenty narrow-petaled pale-blue to white rays around a center yellow disc.

Pearly everlasting

Anaphalis margaritacea

Unlike most plants whose flowers have both male and female parts, pearly everlasting has separate male and female flowers, usually on separate plants. What look like white flower petals are actually papery bracts that surround a yellow-brown disc flower center. The flowers grow in clusters at the tops of the plants and appear to be dried even as they are blooming. Leaves and stems are covered with wooly white hairs.

White rattlesnake-root

Prenanthes alba

You might find white rattlesnake-root growing in the shade of trees in late summer. Its clusters of pinkish-white flowers open into little bells. What look like petals on the dangling flowers are actually ray flowers. White rattlesnake-root can grow up to five feet tall.

FIELD NOTES STALKING THE DOWNY RATTLESNAKE-PLANTAIN

Throughout the summer we've made several trips to Falls Creek Scientific and Natural Area, hoping to catch downy rattlesnake-plantain in bloom. Several small, ground-hugging colonies grow here, their leaves marked with shiny silvery patterns. Only a few plants in each colony might bloom each season, and we've been tracking the progress of the small budded stalks, but we've never captured downy rattlesnake-plantain when it wasn't either in bud or hadn't already gone to seed.

The trees at Falls Creek have started to turn, and light falls through the green, gold, and occasional red leaves like luminous rain. We hike over the wrinkled terrain past pines where we've seen lily-leaved twayblade blooming earlier in the summer, the plants' leaves tattered and limp now, seedpods rattly and dry. When we come to a small patch of downy rattlesnake-plantain near a creek where water slips over rocks, we look eagerly, but the flowers have already gone to seed.

So, we climb up again through more tall trees and across open patches of prairie where goldenrods and asters bloom.

Maybe next year we'll catch the elusive flowers blooming. For now, we're grateful for golden places, wild with color and light.

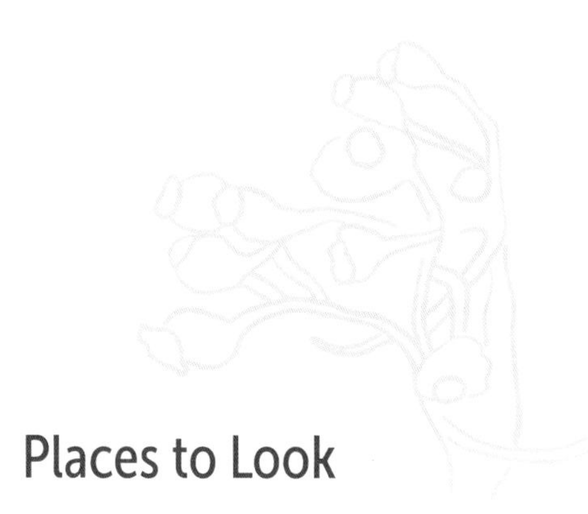

Places to Look

You might have visited **Baker Park Reserve** near Orono in the spring for spring ephemerals, including trout lilies, but come back for the fall forest colors and the seeds and berries under the trees.

Carleton College Cowling Arboretum in Northfield has splendid fall foliage overhead, but look, too, on the forest floor for gone-to-seed spring flowers such as Jack-in-the-pulpit.

Whitewater State Park just south of Elba is a mix of hardwood forest and prairie landscapes. Fall trees burn with color, as does the prairie.

Wild River State Park on the St. Croix River has forest flowers in fruit as well as leaves turning bright overhead.

Maplewood State Park near Pelican Rapids, where prairie meets forest, is a good place to look for spring wildflowers gone to seed, but you can also soak up the colors of the late prairie and of the forest in the fall.

Interstate State Park by Taylors Falls has wooded hills and bluffs overlooking the St. Croix River. Look for large-leaved aster under the trees.

Any of the woodlands where you found spring wildflowers will be a great place to look in the fall as well for flowers gone to seed and berries. As a bonus, the leaves overhead will be spectacular.

AUTUMN

PRAIRIE FALL

Through the Tall Grass

Even into autumn, flowers persist in the prairie, providing food for native bees, birds, and other animals. Gentians, goldenrods, asters (including our favorite, heath aster) brighten the prairie. More than half the plant life in a prairie (by mass) consists of grasses, and now these stand out in their glory, seeds shining in autumn light. Many of the flowers have already died back, the plants' energy settling into their deep roots for the coming winter. Others are setting seed for next year's new plants. Wind, birds, wild animals (and hapless wildflower seekers at times) help disperse the seeds that, along with the deep underground roots, will bring the prairie back again come spring.

Glory in the Grasses

Big bluestem

Andropogon gerardii

Big bluestem is a grass, but it is a true prairie plant, its long stems (up to seven feet tall) bending in the wind. Several flower spikes at the tops of the stem are covered with pairs of tiny flower clusters. If you separate the stalks out, you can see why one of the common names for this plant is turkey foot—the stalks look similar to a bird's track in the mud. The roots of big bluestem may grow ten feet or more deep in the prairie soil.

Bottle gentian

Gentiana andrewsii

Bottle gentian flowers never open. These deeply blue flowers are so tightly closed that large bumblebees are almost the only pollinators strong enough to pry the flowers open and fight their way in. Even then, the bumblebees often crawl only partway inside, using their hind legs to hold the flower open so they can back out again—an amazing sight to see in the wetter parts of the prairie.

Downy gentian

Gentiana puberulenta

Downy gentian brightens the fall prairie with its characteristic gentian blue. Flowers have five spreading lobes and grow in a cluster at the top of the plant to look like lovely upward-facing bells. The sepals on the outside of the petals are a rusty purple color. The "downy" part of its name comes from minute white hairs on the stem and bases of the leaves. While some gentians prefer the wetter places, you might find this gentian growing in drier parts of the prairie.

Grass-leaved goldenrod

Euthamia graminifolia

From zigzag goldenrod to bog goldenrod to stiff goldenrod, the list of goldenrods that grow in Minnesota is long, but this is one of our favorites. True to their name, the leaves are like small blades of grass, about one-eighth inch wide, along the stems. The yellow flowers grow in flat clusters at the tips of stems up to four feet tall, and this plant is sometimes called flat-topped goldenrod.

Greater fringed gentian

Gentianopsis crinita

Greater fringed gentian flowers grow one to a stalk, and the edges of the four deep blue flower petals are fringed or frayed. Look for greater fringed gentian in wetter places in undisturbed prairie with its distinctive fringed petal edges and single deep blue flower.

Great Plains ladies'-tresses

Spiranthes magnicamporum

When we first saw these small, white orchids growing in a prairie, we thought they were nodding ladies'-tresses. One guidebook, though, said to smell them—Great Plains ladies'-tresses orchids smell like almond, and so did these. Look, too, at whether the plant has leaves. Great Plains ladies'-tresses' leaves wither away before the plant flowers.

Hairy grama

Bouteloua hirsuta

The tiny flower clusters of this prairie grass grow along the top of the stalk like fuzzy caterpillars crawling. (Members of the grama family, such as side oats grama, often grow on only one side of the flower stalk.) You might find hairy grama growing in sandy or rocky places. Wherever we find it, its one-sided appearance makes us smile.

Heath aster

Symphyotrichum ericoides

Heath aster is made up of dense clusters of small, white flowers with reddish-yellow centers, sometimes growing only on one side of a stalk. The clusters of flowers crowded by small bristly leaves help to identify this one of Minnesota's many asters in the fall prairie.

Indian grass

Sorghastrum nutans

You can disappear from view in among Indian grasses, which grow to seven feet tall. The flower heads at the tops of stems are long, dense, and yellow-brown, and look golden in the sun. Although we don't like to interfere with any native plants, running a hand along the seed heads is almost irresistible. But appreciating their golden stalks waving in the wind is almost as enjoyable.

Little bluestem

Schizachyrium scoparium

Although flowering plants might be what catch our attention first, prairies are actually composed of roughly 80 percent grasses and 20 percent flowers by sheer mass alone. Little bluestem is a bunchgrass, growing in clumps up to three feet high. In the fall, the plants take on a reddish-golden hue, and the tiny flower clusters at the tips of the stems have a feathery appearance as they go to seed.

Panicled aster

Symphyotrichum lanceolatum

Among Minnesota's many species of aster, this one is the most common white-flowered aster. Its flowers have twenty or more white petals (ray flowers) in loose clusters at the tops of stems. Whereas heath asters prefer drier prairie places, panicled asters often grow in wetter parts of prairies.

Stiff gentian

Gentianella quinquefolia

Usually less than a foot tall, stiff gentian has clusters of purple flowers at the tips of its branches. The tops of the petals fold in to almost close off the insides of the flowers, making the flowers look like clusters of little pointed bottle rockets waiting to be fired.

Swamp lousewort

Pedicularis lanceolata

The yellow flowers that grow sideways on the flower spike may look at first like wood betony, but swamp lousewort is larger and its leaves are not as deeply lobed. The best way to tell the difference between the two plants? Wood betony blooms in the springtime in woods and prairies while swamp lousewort is a fall-blooming plant.

Thimbleweed

Anemone cylindrica

With its whorl of leaflike bracts halfway up its stem, thimbleweed looks as though it is wearing a leafy skirt. The center of the flower is shaped like a thimble surrounded by five white petallike sepals. When thimbleweed goes to seed, the "thimbles" lengthen and turn fluffy and puffy, with tiny brown seeds attached to cottony tufts.

Milkweed

Rattlesnake master

Round-headed bush clover

SEEDS ON THE MOVE

For years, unless a plant was flowering, we found ourselves helpless to identify what that plant might be. Gradually, we learned to recognize some of the distinctive seeds that flowers ripen into (at times from picking the seeds out of our socks, particularly a plant we nicknamed "nasty grass" for its sharp awls that burrowed into our skin).

Here are a few plants whose seeds make them standouts in the fall prairie:

Milkweed plants' seedpods burst with small brown seeds that ride the wind on silky tufts.

The seed heads of round-headed bush clover really are round-headed.

Some plants have single seeds, some have umbrellas of seeds, some form pods packed with seeds. Some seeds hitch rides on fur or feathers or in the guts of animals who eat them. Some simply fall to the ground. However seeds spread, all are engaged in making more of themselves, of guaranteeing next year's prairie.

Gray-headed coneflower

FIELD NOTES

GENTIAN HAT TRICK

The moon sails down the western sky and the sun climbs in the east as we head down to Iron Horse Prairie. Sunshine, a brisk breeze, blue skies—a perfect prairie day.

The grasses are turning golden and red. Scattered among them, glowing like luminous Easter eggs, we spot cluster after cluster of vivid blue bottle gentians.

Wandering from find to find ("Look! Here's another!"), we notice a smaller blue flower growing in single blooms on stems, like tiny iris buds just opening—greater fringed gentian. Farther on, at least thirty different stiff gentian plants grow with their tiny pale flowers opening in clusters like little cups.

Downy gentian

The guidebook we're using lists four kinds of blue gentians in Minnesota, and we've seen three within half an hour. As a bonus, what looks to be a nodding ladies'-tresses orchid smells strongly of almond—a sure sign it's a Great Plains ladies'-tresses instead.

A new orchid and a gentian hat trick—even as flowers set seed and fall settles in, the prairie still holds surprises.

Later, on our way back home, we stop at another prairie on the chance that we might see downy gentian, a fourth kind of blue gentian. We do find them, along with still more stiff gentians, all a deep and lovely blue as the light through the grasses slants toward sunset. A gentian hat trick plus one.

Places to Look

Carpenter Nature Center near Hastings has restored prairie where prairie grasses grow along with coneflowers, goldenrods, blazing stars, wild bergamot, spotted Joe-pye weed, golden Alexanders, and asters. Follow one of the paths to an abandoned cemetery where prairie flowers bloom.

Sixth Avenue Greenway in Minneapolis has planted prairie flowers and grasses. Look for stiff goldenrod, asters, butterfly-weed, rough blazing star, little bluestem, and prairie coreopsis, along with other prairie plants, some gone to seed and some still blooming.

Glacial Lakes State Park near Starbuck preserves native prairie where you might find big and little bluestem, Indian grass, prairie clovers, coneflowers, and goldenrods—blooming or in seed, depending on when you visit.

Forestville State Park near Wykoff is a meeting place for prairie and woodland where spring-fed streams flow out of the rocks. In the prairie, look for compass plant, milkweeds, asters, rattlesnake master, round-headed bush clover, stiff gentian, and rough blazing star. Fall forest foliage is spectacular.

Pipestone National Monument near Pipestone is the site of the Red Pipestone quarries and home to a prairie remnant where more than five hundred species of plants have been cataloged.

Upper Sioux Agency State Park south of Granite Falls has prairie where you can see milkweeds, coneflowers, goldenrods, wild bergamot, blue vervain, and other prairie grasses and flowers.

Go to any of the places you might have visited for the spring or summer prairie to see autumn bloomers and grasses and flowers gone to seed.

EPILOGUE

Smooth sawgrass

TURNING OF THE SEASONS

The Peace of Wild Places

Now the land—forest, prairie, lake, bog, marsh—rests and waits for spring.

And so do we.

But we still plan for the next year of searching for wildflowers.

We started the year with ephemerals, but in a way, all landscapes are ephemeral. This doesn't mean we shouldn't protect them—our native flowers and landscapes are irreplaceable and deserve our best efforts to conserve them. They are also dynamic places continually shaped by blowing wind, flowing water, sun, rain. We humans, too, have an enormous impact on the land. The land left alone perpetuates itself and its communities, but the communities also change.

Whenever we visit a familiar place, we have some idea of what we might see, but we are often surprised at what is or isn't blooming this year or at this time. So, we go in hope of what we might see. And we go, too, for the peace of wild places and the comfort we find there.

* * *

From a distance, McKnight Prairie, part of the Carleton College Cowling Arboretum, looks like two green hills rising from the farm fields around it. Close up, it's chock-full of native plants and near enough to the Twin Cities that we visit several times a year. This was our first visit this year, and the west hill, closest to the road, looked a vivid green as we drove up, while the east hill varied from green to brown. As soon as we stepped onto the prairie we knew why: the west hill had been burned since our last visit, taking away all last year's dried grasses and plants and leaving charred sticks of sumac.

McKnight Prairie

cream wild indigo plants in flower—the first time we've seen so many at once.

Crossing from the burned to the unburned part sped up the season. At the edge of last year's dried grasses yarrow bloomed, and on the east unburned hill we found large-flowered beard-tongue in bud and ground plum, fringed puccoon, hoary puccoon, long-bracted spiderwort, and white camas all in flower. Prairie smoke and kittentails had already gone to seed, flower buds covered the plains prickly pear cactus, and butterfly-weed was just starting to turn orange.

Fire renews the prairie, adds nutrients to the soil, and takes down brush and trees that would turn the prairie into woods if given enough time. We saw the leaves of many familiar prairie plants emerging from the ground that crunched under our feet: field pussytoes, blazing star, goldenrod, compass plant, leadplant. We identified tiny round flowers of prairie blue-eyed grass and saw bastard toadflax and golden Alexanders in bloom. On the far side of the hill, bunches of yellow blossoms marked the sites of almost a dozen

I went to the prairie mourning a friend who had just died, and the prairie helped me remember that birth and death and all of us are part of the same circle. I came home comforted by birdsong, wind, wildflowers, and new green plants under the wide prairie sky.

Field pussytoes

Violet wood sorrel

Puccoon

GLOSSARY

ALTERNATE LEAVES leaves that are not opposite each other along a stem (e.g., black-eyed Susan)

BASAL at the base of

BASAL LEAF a leaf at ground level on a plant at the base of the stem (e.g., downy rattlesnake-plantain)

BIOME a large ecological environment made up of related habitats or communities of plants and animals

BLADDER a hollow sac on a plant such as bladderwort that fills with air and water

BOG a poorly draining acidic wetland made up mainly of peat; the only moisture comes from rain and snow

BOREAL referring to northern parts of the country or the globe

BOTANY the study of plants

BRACT a modified leaflike structure growing just underneath a flower (e.g., Indian paintbrush)

BUD an undeveloped or unopened leaf or flower

CALCAREOUS containing calcium carbonate

CALCIPHILE a plant that thrives in a calcium-containing environment such as a calcareous fen. Edible valerian is an example of a calciphile.

CALYX all of the sepals of a flower together

CARNIVOROUS PLANT a plant that gets at least some of its nutrients by trapping and digesting insects (e.g., purple pitcher plant)

CHLOROPHYLL the green coloring in a plant that converts sunlight, water, and carbon dioxide into food

CIRCUMBOREAL occurring throughout the regions encircling the North Pole

CLASPING LEAF a stalkless leaf that wraps at least partly around a stem (e.g., large-flowered beardtongue)

CLUSTER a group of flowers or leaves (e.g., hoary puccoon)

COMPOSITE FLOWER a cluster of many flowers in a single flower head. Some composite flowers are made of ray flowers (narrow individual petals), some are made of disc flowers (tiny tubular flowers), and some, such as sunflowers, are made of both ray flowers and disc flowers.

COMPOUND LEAF a leaf made up of two or more smaller leaflets (e.g., leadplant)

COROLLA all the petals of a flower together

DECIDUOUS losing its leaves in fall and growing new ones the following year

DISC FLOWER one of many tiny tubular flowers that together form the center of the flower heads of composite flowers such as sunflowers or asters

ECOSYSTEM a community of interacting and interdependent plants, animals, and their environment

ENDANGERED a species that might become extinct in all or a significant number of the areas in which it grows. Ram's head lady's-slipper is endangered in Minnesota.

ENDEMIC a species that has evolved and grows only in a unique habitat or location and nowhere else. Dwarf trout lilies are endemic to Minnesota.

EPHEMERAL lasting only a short time. A spring ephemeral is a flower that grows quickly in the forest before trees leaf out fully, then dies back and disappears completely when the tree canopy shades the ground. Snow trilliums are an example of a spring ephemeral.

EVERGREEN having leaves or needles that stay on the plant and stay green all winter

FAMILY a related group of plants that have common biological features and evolutionary history

FEN a permanent wetland that has a supply of nutrient-rich groundwater

FLOWER HEAD a dense group of usually unstalked flowers (e.g., wild bergamot)

GRASS a plant with jointed stems, narrow leaves, and seedlike fruit, such as big bluestem or little bluestem

GROUNDWATER water from rain or snow that collects under the surface of the ground. In fens, groundwater wells up out of the earth.

HABITAT all the environmental conditions, such as light, moisture, soil, plants, and animals, in which a plant naturally grows or an animal lives

INFLORESCENCE the entire flower cluster of a plant

INTRODUCED PLANT a plant, often from another country or continent, brought into a habitat where it did not evolve

LANCEOLATE lance-shaped, tapering to a point (e.g., lance-leaved violet)

LARVA (plural, LARVAE) a wingless stage in insects after they hatch. The larvae of moths and butterflies are called caterpillars.

LEAF AXIL the inside angle where the leaf joins the stem

LEAFLET a leaflike division of a compound leaf

LIP an enlarged or modified petal, especially in orchids

LIP ONE'S WADERS to have water come over the tops of your boots

LOBE one of the divisions of a single leaf

MARSH a nutrient-rich, open wetland habitat dominated mainly by grasses

MESIC a habitat or prairie that is neither wet nor dry but receives in-between amounts of moisture

METAMORPHOSE to change completely as when a caterpillar turns into a butterfly

MICROHABITAT a smaller area within a surrounding habitat where specialized plants have adapted to grow. Some rock crevices along Lake Superior form a microhabitat for butterwort.

NATIVE PLANT a species found in an area where it evolved and that has adapted to the local conditions

NODE the point on a stem or a rhizome from which a leaf, a branch, or a new stem grows

OPPOSITE two leaves or flowers growing from a single node on a stem (e.g., bottle gentian)

PANICLE a branched cluster of flowers (e.g., panicled bluebells)

PARASITIC getting nutrients from another plant instead of making its own food. Clustered broomrape is a parasitic plant.

PEAT the buildup of partly decomposed sphagnum mosses and other plants, especially in bogs

PETAL part of a flower, usually the most brightly colored part

PHENOLOGY the study of life cycles in the natural world

PHOTOSYNTHESIS the process that green plants use to convert carbon dioxide, water, and sunlight into food

POLLEN a fine powdery substance in a flower usually necessary for seed production

PUBESCENT hairy

RACEME a single flower stem with many stalked flowers on it (e.g., white wild indigo)

RAY FLOWER in a composite flower, such as a sunflower, one of the individual flowers that looks like a single petal and surrounds the central disc

RELICT a group of organisms, such as a species of plant, that survives in an area where it was once much more widely spread (e.g., butterwort)

RHIZOME a rootlike stem growing horizontally underground

ROSETTE a circular cluster of leaves at ground level (e.g., kittentails)

ROOT the part of a plant below ground that holds the plant in the soil and absorbs water and nutrients

SEDGE a grasslike marsh plant with a three-sided stem. "Sedges have edges."

SEPAL a leaflike part that helps protect the flower bud, usually found at the base of the flower once the bud opens. The number of sepals usually equals the number of petals. Sepals are often green but are sometimes brightly colored and look like petals.

SPADIX a thick spike covered with tiny flowers, often protected by a spathe, as in wild calla

SPATHE the hoodlike structure that shields the spadix or central flower spike in flowers such as Jack-in-the-pulpit and skunk cabbage

SPECIAL CONCERN a species is of special concern when it has become very uncommon or has such highly specific habitat needs that it could become threatened or endangered. Blanketflower is a species of special concern in Minnesota.

SPECIES a population of plants or animals that can interbreed with each other

SPHAGNUM MOSSES a group of more than 380 species of mosses that can hold up to twenty times their weight in water or more.

SPIKE an elongated flower cluster with stalkless flowers directly attached to the central stem (e.g., pale-spike lobelia)

STALK a part of the plant that branches off of the main stem and functions as a point of attachment for a leaf, flower, or fruit

STAMEN the pollen-producing part of a flower

STEM the main part of the plant that supports the leaves and flowers

SWAMP a wetland with woods and standing water that is higher in nutrients and less acidic than a bog

TANNIN a naturally occurring acid in some plants (it can color your socks brown when you wade through tannin-rich water)

TEPAL the term for sepals and petals that look very much alike

THREATENED a threatened species is one that has a good chance of becoming endangered. Kittentails is a threatened species in Minnesota.

TOOTHED having sharply angled or jagged edges, such as cutleaf toothwort leaves or the leaflets of golden Alexanders

TUBERCLE a small enlargement, such as the bump on the lower lip of a tubercled rein orchid

TUBULAR tube-shaped

UMBEL a flower cluster in which all the flower stalks originate from the same point. An umbel can be flat-topped, rounded like an umbrella, spherical, or unevenly shaped (e.g., golden Alexanders).

WHORL a group of three or more leaves growing from the same node on a stem and often circling the stem. Spotted Joe-pye weed has whorled leaves.

RESOURCES

MINNESOTA WILDFLOWERS website is a great online resource for identifying flowers by color, name, or bloom time: https://www.minnesotawildflowers.info.

Find out more about **MINNESOTA'S STATE PARKS,** including directions and what you might see and do at each one, at www.dnr.state.mn.us/state_parks.

MINNESOTA'S SCIENTIFIC AND NATURAL AREAS are places protected by the state that contain native plant communities, rare species, and special geological features. These sites do not have any facilities, staff, or maintained trails, but we love their wildness and the thrill of finding out what blooms in them. For information on where they are located, visiting guidelines, or to get a permit, go to www.dnr.state.mn.us/snas.

THE NATURE CONSERVANCY sites are protected places we like to visit. These sites do not have maintained trails, staff, or any facilities. Before you go, read their visiting guidelines, which are posted along with site locations at https://www.nature.org/ourinitiatives/regions/northamerica/unitedstates/minnesota.

BLUE FLOWERS

Here is a color guide to Minnesota's wildflowers to help narrow your search. But remember, colors also vary. For example, if you don't find the flower you're looking to identify in the pink section of this guide, try looking in the blue section or the purple or the violet.

SPRING

Panicled bluebells
p. 67

Prairie blue-eyed grass p. 52

Spreading Jacob's ladder p. 33

Virginia bluebells
p. 35

SUMMER

Blue flag
p. 83

Blue vervain
p. 83

Great blue lobelia
p. 84

Harebell
p. 65

Marsh skullcap
p. 123

Pale-spike lobelia
p. 103

AUTUMN

Bottle gentian
p. 154

Downy gentian
p. 155

Greater fringed gentian p. 157

Large-leaved aster
p. 147

Stiff gentian
p. 163

GREEN FLOWERS

SPRING

Jack-in-the-pulpit
p. 26

ORANGE FLOWERS

SPRING

Hairy puccoon
p. 48

Hoary puccoon
p. 49

SUMMER

Butterfly-weed
p. 98

Michigan lily
p. 103

Wood lily
p. 112

PINK FLOWERS

SPRING

Bog rosemary p. 132

Fairy-slipper orchid p. 134

Jeweled shooting star p. 26

Prairie smoke p. 53

Rose twisted-stalk p. 76

Rue-anemone p. 31

Virginia spring beauty p. 21

Wild geranium p. 37

SUMMER

Bog laurel p. 132

Common milkweed p. 99

Cross-leaved milkwort p. 83

Dragon's-mouth p. 133

Pipsissewa p. 75

Prairie rose p. 106

Spotted Joe-pye weed p. 88

Showy lady's-slipper p. 77

Small cranberry p. 138

Stemless lady's-slipper p. 77

Sullivant's milkweed p. 108

Swamp milkweed p. 89

Tuberous grass-pink p. 138

Water smartweed p. 124

PURPLE FLOWERS

SPRING

Large-flowered beardtongue p. 51

Long-bracted spiderwort p.51

Purple clematis p. 66

Ram's head orchid p. 86

SUMMER

Butterwort p. 63

Leadplant p. 102

Narrow-leaved purple coneflower p. 103

Prairie blazing star p. 105

Purple prairie clover p. 107

Rough blazing star p. 108

Wild bergamot p. 111

RED FLOWERS

SPRING

Canadian wild ginger p. 24

Red columbine p. 30

Skunk cabbage p. 32

SUMMER

Purple pitcher plant p. 136

Round-leaved sundew p. 137

VIOLET FLOWERS

SPRING

Bird's-eye primrose p. 62

Bird's foot violet p. 44

Gaywings p. 74

Ground plum p. 48

Pasqueflower p. 51

Prairie violet p. 54

Round-lobed hepatica p. 31

Sharp-lobed hepatica p. 32

Showy orchis p. 32

Violet wood sorrel p. 56

Wild blue phlox p. 36

SUMMER

Lance-leaved violet p. 85

Lily-leaved twayblade p. 74

WHITE FLOWERS

SPRING

Bastard toadflax p. 44

Bloodroot p. 18

Bunchberry p. 62

Canada mayflower p. 64

Clustered broomrape p. 45

Cutleaf toothwort p. 18

Drooping trillium p. 24

Dutchman's breeches p. 18

Dwarf trout lily p. 19

Eastern false rue-anemone p. 20

Edible valerian p. 46

False Solomon's seal p. 25

Field pussytoes p. 46

Goldthread p. 74

Large-flowered trillium p. 28

Mayapple p. 28

Nodding trillium p. 29

Seneca snakeroot p. 54

Small white lady's-slipper p. 54

Snow trillium p. 21

Starflower p. 77

Starry false Solomon's seal p. 55

Three-leaved false Solomon's seal p. 138

Twinleaf p. 34

Two-leaved miterwort p. 34

White trout lily p. 21

Wild sarsaparilla p. 67

Wood anemone p. 37

SUMMER

American white waterlily p. 122

Broad-leaved arrowhead p. 122

Buckbean p. 132

Culver's root p. 100

Flowering spurge p. 100

Labrador tea p. 135

Leatherleaf p. 135

Nodding ladies'-tresses p.85

One-sided pyrola p. 86

Ragged fringed orchid p. 86

Rattlesnake master p. 107

Round-headed bush clover p. 108

Tussock cottongrass p. 139

Western prairie fringed orchid p. 109

White camas p. 110

White prairie clover p. 111

White wild indigo p. 111

Wild calla p. 124

Wild quinine p. 112

Yarrow p. 113

AUTUMN

Downy rattlesnake-plantain p. 147

Great Plains ladies'-tresses p. 158

Heath aster p. 160

Panicled aster p. 162

Pearly everlasting p. 148

Thimbleweed p. 164

White rattlesnake-root p. 149

YELLOW FLOWERS

SPRING

Beach heather
p. 44

Bluebead lily
p. 62

Downy painted cup
p. 46

Downy yellow violet p. 24

Fringed puccoon
p. 47

Giant Solomon's seal p. 26

Golden Alexanders
p. 48

Kittentails
p. 50

Large-flowered bellwort p. 27

Marsh marigold
p. 28

Prairie alumroot
p. 52

Prairie buttercup
p. 52

Sessile-leaf bellwort p. 31

Wood betony
p. 57

Yellow star-grass
p. 57

Yellow trout lily
p. 22

SUMMER

Black-eyed Susan p. 97

Blanketflower p. 98

Brittle prickly pear p. 98

Common bladderwort p. 122

Compass plant p. 100

Gray-headed coneflower p. 101

Indian paintbrush p. 102

Large yellow lady's-slipper p. 85

Plains prickly pear p. 104

Prairie coreopsis p. 105

Prairie ragwort p. 105

Prairie sunflower p. 106

Small yellow lady's-slipper p. 87

Tubercled rein orchid p. 90

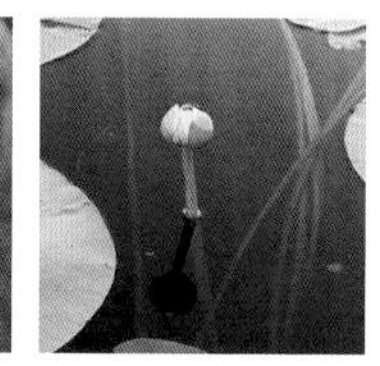
Yellow pond-lily p. 124

AUTUMN

Grass-leaved goldenrod p. 156

Swamp lousewort p. 163

INDEX

A

B

C

D

E

F

G

H

I

J

K

L

M

N

O

P

Q

R

S

T

U

V

W

Y

Z

Phyllis Root is a writer, **Kelly Povo** is a photographer, and they love searching for, learning about, and finding Minnesota's native flowers. Their first wildflower adventure together was more than ten years ago, to **Big Bog State Recreation Area**, where they walked on the boardwalk a mile into the bog and saw plants they had never seen before. Since then, they have waded rivers, slipped down snowy hillsides, and visited more places than they can remember.

Phyllis, an award-winning author, has written many picture books about Minnesota, including *Big Belching Bog, Plant a Pocket of Prairie,* and *One North Star,* all published by the University of Minnesota Press. Kelly, a professional photographer for more than thirty years, has exhibited in galleries and art shows across the country. Her photographs have been published on posters, calendars, note cards, and in books. Kelly and Phyllis have collaborated on several books, including *Girlfriend Gumbo* and *Gladys on the Go*. Whatever they are doing, Phyllis and Kelly laugh—a lot.